SAVING SNAKES

SAVING SNAKES

Snakes and the Evolution of a Field Naturalist

NICOLETTE L. CAGLE

UNIVERSITY OF VIRGINIA PRESS

Charlottesville and London

University of Virginia Press
© 2023 by the Rector and Visitors of the University of Virginia
All rights reserved

First published 2023

9 8 7 6 5 4 3 2 1

Library of Congress Cataloging-in-Publication Data

Names: Cagle, Nicolette Lynn Flocca, author.
Title: Saving snakes : snakes and the evolution of a field naturalist /
    Nicolette L. Cagle.
Description: Charlottesville : University of Virginia Press, 2022. | Includes
    bibliographical references and index.
Identifiers: LCCN 2022026299 (print) | LCCN 2022026300 (ebook) |
    ISBN 9780813948829 (paperback) | ISBN 9780813948836
    (ebook)
Subjects: LCSH: Cagle, Nicolette Lynn Flocca, 1980– | Women naturalists—
    Biography. | Snakes—Research.
Classification: LCC QH31.C265 A3 2022 (print) | LCC QH31.C265 (ebook) |
    DDC 597.96072—dc23/eng/20220817
LC record available at https://lccn.loc.gov/2022026299
LC ebook record available at https://lccn.loc.gov/2022026300

Cover photo: Grass Snake. (vvictory/istock.com)

To my son and future generations of students and snakes

CONTENTS

AUTHOR'S NOTE

Many of the events described are based on the author's memories and experience. However, the names and identifying details of some people have been changed. In these cases, any resemblance to persons living or dead resulting from those changes are coincidental and unintentional.

Excerpts within the chapter "Dissertation" have been modified and reprinted with permission from *Voices for Biodiversity*.

A note on species names. In this book, I have capitalized accepted common names that identify a particular species or subspecies, for example, Northern Red Oak or Common Garter Snake. Names that refer to a group that might include many species are not capitalized, for example, oaks, garter snakes. In some cases, but not all, scientific names of snake species are included parenthetically for clarity.

SAVING SNAKES

AN ACRE OF SNAKES

 THE MILKWEED'S BROAD LEAVES still had the bright-green hue of new growth. I would harvest leaves from those very plants someday to rear the tiger-striped cat-erpillars of Monarch Butterflies, but that would be years later, a project for my early teens.

This day, I was seven, walking through an old field in northern Il-linois with my father. We wandered around a forgotten acre, wedged between a busy road, a tollway, and a suburban neighborhood domi-nated by 1950s ranch houses with low-sloped roofs. A roadside ditch filled with the hotdog-shaped flower heads of cattails screened our view of passing cars. The tall grass tickled my nose and shoulders, but Dad tamped down a path with his sturdy work boots, and I was able to push through. Dad found an old piece of plywood, half-rotten on the ground. His eyes filled with anticipation, lighting up like the clear blue sky overhead, and he leaned down to lift up the plywood board and peer beneath it. Nothing.

We wandered on, toward the back of the undeveloped property where the deer had flattened the grass for a summer nap. Dad saw another piece of plywood. His eyes shined again as he leaned down to lift the board. Underneath lay a garter snake, outstretched to reveal rough, gray scales overlain with faded gold stripes. The snake was beautiful, like one of those golden bracelets from ancient Rome I had seen at the Field Museum in the city, and it made my breath hitch. The outstretched garter snake was powerful, filling me with inchoate awareness of my own potential—potential to discover hidden beauty, to explore unknown places, and to teem with anticipation again and again.

That garter snake also embodied a useful capacity for adaptation: it was well-suited to living beneath that crumbling plywood board. It was a perfect spot to hide from predators, like raccoons, hawks, and even other snakes. It was also a perfect spot to find food, mostly little invertebrates that like dark, wet places: slugs, snails, and spiders. Although, truth be told, garter snakes will eat almost anything that moves, including frogs, small birds, baby rodents, and even other little snakes.

Garter snakes don't need much room—they can make a life on an acre of forgotten field. One researcher recently calculated the home range of garter snakes from a nearby population at between 0.015 and 0.14 acres, an area roughly as big as the end zone of an American football field, and they move just three to fifty-seven feet all summer. That is, these snakes stay pretty close to home, especially on clear summer days when fathers and daughters are out looking for them.

Each time Dad and I visited the little refugium of undeveloped land, we found only a few garter snakes, but it is likely that a dozen or more lived there. The highest density of garter snakes on record comes from a remnant prairie in Ohio, where up to thirty-five garter snakes were found per acre.

On our walks over the years, my father and I found many snakes in that old field, more garter snakes and Dekay's Brown Snakes too. Those snakes were companions as we explored the path to the lake, and another to the river, and another to the sugar maple woods. Those snakes offered excitement on dull days, awe during otherwise uninspired moments, and beauty when the world around us could be so ugly. Those snakes solidified bonds between father and daughter, child and place, and domesticated human existence and the wild animal realm.

When I was in high school, the little acre was scheduled for development. My mother wrote letters of protest, polite but fervent. She described the value of that old field to butterflies and snakes and deer and children. Development stalled for a while. We thought we had won a conservation victory.

But a few years later, about the time I left for college, that little acre of old field—the patch of verdant Monarch habitat with its generous snakes—was bulldozed. The rich, black earth was pounded, compacted, and covered in asphalt. Developers erected six identical townhomes with faux-brick fronts and views of the tollway overpass.

I cried when I saw the development for the first time. I mourned the creatures. I mourned the young garter snakes that would never again call that acre home. I mourned the green space stolen from a neighborhood of kids. I mourned the memories that I would never make there with my father or mother or spouse or child or friends.

That was the first time I cried for snakes, but not the last. Across the globe snakes are dying. Severe declines in snake populations have been recorded in North America, Europe, Australia, and Africa. Data suggests that for some of these populations, those steep declines started only around the turn of the twenty-first century, but some began decades earlier. The threats to snakes are widespread and human-mediated: habitat loss and fragmentation, road construction and use, and disease. Plus, conservation efforts for snakes lag behind those for other taxa. It is difficult to motivate people to protect creatures they consider cursed or creepy, and recent research suggests that fear of snakes may be at least partially innate.

That innate fear, however, has not stopped other cultures from revering snakes, associating them with healing, wisdom, and creation itself.

The Hopi, an Indigenous tribe in the southwestern United States, have a long history of reverence for snakes. Every other August, on the mesas of the high desert, an intimate ritual with snakes conjures agrarian fecundity. For over a week, preparations are made for the final snake ceremony meant to encourage rainfall and the maturity of traditional crops, such as corn, squash, and beans. In preparation for the final Snake Dance, snakes are collected over the course of four mornings from the four directions and then purified by the snake priest using specially prepared water. Songs are sung to the snakes until Orion rises in the east.

The ceremony culminates with the Day of the Snakes, where up to one hundred snakes, including rattlesnakes, are corralled by cottonwood limbs and dancing begins. One person from the tribe carries a snake, and there are photos from the 1890s with a Snake Dancer gently holding the snake between his teeth. Black-and-white footage of the Snake Dance from the early 1900s shows men dressed in traditional regalia. At first, the men step to the rhythm of a drum, forming a perpetual motion circle in the center of the crowd. Each man disappears through a thick stone arch and then a new man seems to materialize, giving the impression of an infinite number people joining the circle, emerging from an otherworldly realm. Later, the men form lines. Each person leans forward, taking steps in place that raise the fringe on ankle-high moccasins, moving to a sacred beat that the silent film refuses to share.

After the dancing has finished, the snakes are released to the four directions, carrying prayers for rain and plenty, serving as messengers to the spirits. In 1913, President Theodore Roosevelt observed the Snake Dance and noted that "many of the tourists did not show the proper respect" for the ceremony. Hopi elders shared those sentiments, and today this ceremony of gratitude and good fortune is largely closed to non-Indigenous visitors.

Martin Nillson, an early twentieth-century scholar of ancient Greek and Roman religions, described the importance of snakes in domestic worship in ancient Greece. There the snake symbolized the house god, a form of Zeus, and it was given offerings of water, oil, and fruit in amphorae. In Sparta, the Dioscuri—or twin sons of Zeus, Castor and Pollux—were seen as protectors of the home and were associated with depictions of snakes. In fact, snakes often lived in houses among the ancients, tamed by good care and feeding. In Minoa, snakes in one's home were sometimes called the *Dios kouroi,* or sons of Zeus.

Similarly, Nillson wrote that in the Balkans and Greece in the early twentieth century, snakes found in one's home were welcomed as the "guardian spirit of our house" and greeted as a grand lady might be. In 1940, Nillson recorded that the veneration of the house snake still occurred in Europe. He described a man he knew personally in his own country, Sweden, who offered milk to living house snakes.

I count myself among these people—those who revere snakes for their mystery and history, those who have a soft spot for the underdog, and those who can find beauty in unexpected places. For me, inculcation into the cult of Ophidians—another word for snakes that derives from the Greek *ophis*—was rather simple: a few walks of discovery with a comfortable guide and an acre of old field that yielded hidden delights. Perhaps developing respect and appreciation for snakes could be that simple for everyone, if only we gave them the opportunity.

LESSONS FROM WISCONSIN

 I STOOD ON THE beach, my toes curling into the hot, coarse sand, my skin both overheated from the midday sun and overcooled by the lakeside breeze. There were a lot of people along Lake Michigan's shore that day, reveling in the warmth, drawing out those canicular days before the leaves fell and the sand chilled and the water numbed their toes.

The muddy turquoise lake lapped the beach. Puckish waves peaked without breaking, like naughty elves playing at people's waists and knees. Children were scattered across the beach, building sand castles, digging canals, teasing the lake by running in and out, trying to outpace its surges.

I hadn't arrived at the beach alone. Earlier that week, I joined my friend's family—the Steins—for a magical week at a crunchy Wisconsin farm north of Chicago. There, we freed potato plants from the ravages of squishy, orange Potato Beetle larvae by day and listened to the low rumblings of guitars by night. We ate vegetarian meals and took mud baths in the creek. I perfected my downward dog and learned how to meditate. It was a transformational experience; starting a meditation practice at that age probably changed the course of my life. It was also the longest I had ever been away from my parents, and I was ready to get home, but the Steins and I made a stop at the lakeshore as we headed back to Illinois.

I was fourteen, an age when my hair frizzed and curled willfully, my teeth ached under the pressure of metal wires, and my limbs felt too long for my torso. I didn't feel like talking to anyone. I wanted to protect my thoughts, gather them, savor them, and imprint them deep in my mind so that no amount of schooling or churlish gossip or fear could desecrate the profoundly grounded spirituality of that week.

Alone, my feet now deep in the sand, I noticed a little blonde girl, maybe three years old, scampering away from the water's edge. At first I thought she was trying to outrun the waves, but the intentness of her stare up the soft slope of beach made me look harder.

A snake.

The little girl was running toward a snake.

My first thought, quick and irrational, was that I needed to save her. I sprinted across the sand, barely beating the little girl to her object of interest.

It was a Northern Watersnake, coarse and coppery with brown banding, of medium thickness, maybe a foot and a half long. These snakes are harmless, but defensive when poked or grabbed or otherwise needlessly startled.

There was an inexplicable sense of urgency, a powerful protective instinct reaching across eons of evolution. A little girl was going to be bitten. I swooped my too-long arm down and grabbed the snake clumsily by the middle. The startled snake immediately bit me, once and then again. I ran off with it, toward the end of the beach that was least crowded, where the woods started to overtake the sand. I set the snake down unceremoniously and turned around. The little girl was gone. My heroism was null and nugatory.

I began to walk back toward the Steins. I could see they were packing up to go. My hand was bleeding, more than you would expect from such tiny bites. That was probably on account of anticoagulant from the snake's saliva having entered my wounds.

My hand was also starting to go numb, but that wasn't entirely unexpected. I had been bitten by dozens of tiny, baby watersnakes in recent months—I could never get enough of snakes—and I always noticed a bit of numbness at the puncture site.

I kept walking and reached the Stein family. They were almost packed up. I grabbed a blanket that one of them had rolled up, slipped on my loafers, and walked with them back to the car.

As we walked, I talked to my friend's little sister, Rochelle. Rochelle was a few years younger than me. She had smooth brown hair and lively, intelligent eyes. At eleven, those eyes seemed to notice everything—she watched the people around her carefully, amusement playing on her face.

I told Rochelle how I moved a snake away from a little kid and I had been bitten. I tried to make my voice sound nonchalant, but it wavered with excitement. She gave me a sideways glance and smiled wryly. Like the little girl, Rochelle wasn't feeling compelled to give me brownie points for good citizenship.

We climbed into the stuffy car. I rolled down my window by hand. I was in the back seat, but at least I wasn't in the middle. We started the two-hour drive home.

I noticed the numbness in my hand again; the lack of sensation was moving, expanding. My forearm had gone numb too. I started to get uncomfortable and nervous.

I stared out the window. I took deep, calming breaths. The numbness slithered upward. My stomach tightened.

I finally blurted out, "Umm guys . . . my arm is going numb." I didn't provide any context. I hadn't told anyone but Rochelle about the snake.

The Steins needed more information. I was reluctant to tell my story, hampered by an adolescent disinclination to share but compelled by my worry.

When I was done with my story, there was only silence. Everyone was thinking the same thing, but no one wanted to give it voice, except Rochelle: "Are you sure it wasn't poisonous?"

I stiffened. She had voiced my fear. Had I misidentified the snake? Was there some venomous species up in southern Wisconsin that wasn't found in northern Illinois? I was only a teenager; my world was small. Did I make a mistake? Was I going to die?

"It was a Northern Watersnake." I deadpanned my reply. Inner turmoil stabbed my stomach, but I tried not to let on.

"Well, keep us posted," Mr. Stein said, and we kept driving. Midway through the drive, the numbness had reached my shoulder; my entire arm was without feeling, but it stopped there. I monitored myself compulsively. Was the numbness moving? What if it reached my heart? Was I breathing okay?

I made it home. I lived. And I didn't mention a thing to my parents.

I learned a couple things from that episode. First, I needed to always learn the local snakes before traveling somewhere new. Second, if people say things with confidence, other people will believe them, even if the words are coming out of the mouth of a fourteen-year-old.

Since then, I've learned more about Wisconsin's snakes. Wisconsin is pretty far north, and colder climate typically means fewer snakes. Compared to Illinois's thirty-seven, Wisconsin boasts only twenty-two snake species, two of which are venomous and rare: the Eastern Massasauga and the Timber Rattlesnake. Neither of them is found in the region of Wisconsin I was in.

While my emotional reaction to my bites makes a kind of sense, my physical reaction to watersnake bites initially baffled me. What I now know is that many snake species thought to be harmless in the United States, including common colubrid snakes like garter snakes and the Northern Watersnake, have a modified salivary gland, better known as the Duvernoy's gland, which releases secretions. The Duvernoy's gland can be found behind the eye of these snakes, and it is connected to the snakes' mouths via ducts. In some species, the secretions of Duvernoy's gland are potent toxins, but in other species, the toxicity is quite weak. These snakes typically aren't considered venomous because the venom isn't potent or it lacks a clear, toothy delivery-system, but nowadays researchers estimate that 40 percent of colubrids produce some sort of venom. The truth is that some snakes that are completely harmless to us are actually venomous—like Ring-necked Snakes and garter snakes.

With the Northern Watersnake in particular, the rearmost maxillary fangs are a bit larger than its other teeth, and it's common knowledge that wounds from their bite usually bleed more and their saliva is said to be an anticoagulant. Little research had been done on the anticoagulant properties of Northern Watersnake saliva or its possible applications for human medicine, which is strange because snake venoms are widely researched for pharmaceutical properties. Drugs have been developed from snake venoms that reduce blood clots in stroke victims (thank you, Malayan Pit Viper) and might speed the healing rates of cuts (nice one, Australian Taipan).

But in 2010, an impressive undergraduate student from the Pennsylvania State University, Daniel Ranayhossaini, showed that Northern Watersnake saliva displays hemolytic activity, bursting the blood cells of fish. Ranayhossaini postulated that the saliva contains cyclodextrins, rather than the traditional proteins and enzymes of most venoms. Cyclodextrins burst human red blood cells too, and they can trigger allergic reactions. This new research provided a possible explanation for my own mysterious reaction to Northern Watersnake bites, but in the 1990s, I still had a lot to learn.

IDOLS

I FIRST ENCOUNTERED Donald Culross Peattie when I was sixteen. I stood in a long, high-ceilinged room with hardwood floors, portraits on the wall, and carved chairs that flanked the fireplace like wooden thrones.

The walls were lined with bookshelves. I walked slowly down the room, my eyes scanning old books, dusty and worn. I stopped when I saw *A Book of Hours*. I gingerly pulled the blue hardcover book from its home. I laid the spine into my left palm and carefully lifted the cover. I turned a few pages and read my first words of Donald Culross Peattie's oeuvre. The writing reflected deep presence in nature, an appreciation of science that bordered on mysticism, and startling and profound curiosity about everything.

Peattie was born in Chicago in 1898 to talented parents. His father, Robert Burns Peattie, was a journalist, and his mother, Elia Cahill Wilkinson, was a spirited writer and poet. Peattie himself grew up to be a prolific writer of newspaper articles, book reviews, and books. His interests were wide ranging. He waxed philosophical, wrote a history of Vence, France, and was an avid mixologist—I highly recommend his recipe for hot buttered rum, especially on a cold winter's evening. Peattie also trained as a botanist at Harvard and was a careful observer of the natural world. While Peattie the Philosopher and Peattie the Mixologist are fascinating, it was Peattie the Naturalist whom I fell in love with. In *The Road of a Naturalist,* Peattie wrote: "Nature is an ultimate sanctuary for sanity and goodness; American nature is a first national principle. To it I am dedicated." That dedication showed.

Peattie wrote little love songs to the Chicago prairie, regarded by others as no more than a soggy wasteland, explaining that "Prairie as I knew it meant misty distances, and a sweet south wind that brought thaw and the sound of church bells from the Bohemian settlements beyond the groves. It meant these island groves of willow and cottonwood, and the piping of frogs from too far to walk, and shooting-stars that came to flower in the tardy springtime of this mild flat wilderness."

He also wrote little love songs to Chinquapins, small trees related to the American Chestnut, couched in the history of Captain Smith arriving in Virginia and grounded in the realities of timber and wildlife use, saying that the "first mention to be found of the Chinquapin is in Captain John Smith's account of Virginia. But the creatures of that wilderness had known and appreciated it from time immemorial."

And he wrote love songs to planets, including Saturn rising in an "inky violet" sky at 3:00 a.m., a morning planet that "reminds us that other worlds than ours do not keep our hours nor move upon the rounds appointed to us." Peattie always knew how to squeeze the true essence out of natural phenomena, finding meaning even in the vines clinging to a window that eventually blocked his view of Saturn and "so, a leaf on little earth has power to blot it out."

Peattie's writings have taken me to wild and beautiful places. I've been there in real life too. I have made an effort to follow in his footsteps, but I found many of those beautiful places changed by nearly one hundred years of human activities resembling something far less than progress.

One of those places was Vence, where Peattie lived for a while. In the 1920s and 1930s, it was quaint and virtually carless, with trotting donkeys and flowers bursting through a green carpet in spring and autumn. This was a medieval walled village tucked into the steep hills less than twenty kilometers from Nice in the South of France, a quiet, agrarian community that attracted brilliant writers, like D. H. Lawrence and James Baldwin, and world-class visual artists, like Marc Chagall and Henri Matisse. But the truth was Vence always stood in the shadow of the nearby Saint-Paul de Vence, where Baldwin is buried and Chagall died in 1985.

In Peattie's time, Vence was all singing chaffinches and fragrant maquis; it was all luscious olives and weathered stone walls. When I went in 2017 to track down Peattie's home there, called La Roselière, the city had become all soul-sucking new construction and pressing suburban sprawl and sterile swimming pools. The charm of the past was replaced by the call of comfort, the mantra of merchandise, the melody of the Mediterranean McMansion. If one wanted to retrace Peattie's footsteps and walk the three-kilometer route from La Roselière to the center of Vence, one had to run across a busy highway and press against the railing of a bridge to avoid getting plastered by a car. Vence was no longer a sleepy respite from Saint-Paul de Vence; it had become a thoroughfare to get to brighter French Riviera gems, like Cagnes-sur-Mer and Grasse.

Peattie also left his literary mark on the state where I now live, North Carolina. In two of his books, *Flowering Earth* and *The Road of a Naturalist,* it becomes clear that the natural and botanical beauty of his visits to western North Carolina as a child inspired him to be a close observer of nature. At age eight, Peattie "began to discover a world older and greater" in the Blue Ridge Mountains of Tryon, North Carolina.

In fact, Peattie was so enamored with the natural richness of this area that when he got word that a favorite glen of his, complete with "tall trees and a waterfall," was to be sold off to loggers, he wrote another literary love song that inspired a wealthy man to purchase Pearson's Falls and protect it. At Pearson's Falls, Peattie spent "golden hours" cataloguing the flora, fauna, soil, and rock of the place throughout the seasons. Peattie described the "trees with broad, filmy leaves, airy in their bud in spring, cool and shady, but admitting plenty of light in summer, glorious in autumn, naked in winter when the delicate tracery of their boughs and twigs is revealed."

Today, Pearson's Falls is a 268-acre botanical preserve owned by the Tryon Garden Club. It is also a Blue Ridge Natural Heritage Area and part of the North Carolina Birding Trail. Peattie himself noted the preserve's variety of warblers and vireos, sparrows and thrushes, and woodpeckers too. He delighted in the dizzy spirals that the Brown Creeper made as it inspected bark crevices for insects, and he fell into reverie with "the tameless happiness and swift melodies of the warblers."

In contrast to Vence, a visit to Pearson's Falls today is still filled with the numinous enchantment of the natural world. A fresh veil of water still cascades over the old granite rock, which give the ninety-foot falls three peaceful tiers, one tall and dramatic, one flat, and another where the rock slants upward and seems to make the water burble and bubble all the more. The creek still sparkles as light reflects from pebbles of pearly quartz and mica. The preserve is still a rich "little museum" of flora, with a forest floor of violets, Wild Ginger, and Fairy Wand and colonies of moss near the waterfall, a shrub layer of spicebush and hazel, and an overstory of beech and oak.

I would have followed Peattie anywhere, but it probably would have been a bad idea. Sometimes, the thing you love most is the one thing your hero can't abide: snakes. Peattie never wrote literary love songs to snakes, and his writings indicate that he didn't like them much. Plus, the adage "never meet your heroes" exists for reasons beyond, as Louisa May Alcott noted in *Jo's Boys,* the disappointment we often feel "when we discover that our idols are very ordinary men and women." I had been loath to follow that advice, but I shouldn't have ignored the wisdom of ages.

Around 2012, I had the opportunity to meet another famed naturalist, E. O. Wilson. I had seen E. O. Wilson speak in college, and I was a fan of a man who could both practice science and write about it eloquently. Perhaps more importantly, Wilson's illustrious career as a Harvard entomologist and sociobiologist was launched by a precocious boyhood interest in snakes in one of the most snake-diverse regions of North America, the southeastern United States. Wilson's books describe his experiences catching watersnakes down at the pond, building wood and wire-mesh cages, and having dangerous encounters with Cottonmouths. In fact, Wilson described himself as the "snake-wrangler-in-chief" at Boy Scout Camp Pushmataha, earning himself the nickname of "Snake."

The shift from snakes at Camp Pushmataha to ants at Harvard wasn't particularly abrupt. Despite his work as an entomologist, snakes had left an indelible mark on Wilson, and he continued to write about snakes in his books *The Naturalist, Letters to a Young Scientist,* and *In Search of Nature.* In the latter, Wilson reflected on the fear and fascination snakes inspire among people around the world, and moved into sociobiological theory to explain why this may be. The idea is a familiar one, although contested by authors like Jared Diamond: for most of human history, we lived in hunter-gatherer bands with close proximity to snakes, many of which were dangerous, and like other primates, we developed a genetically subsidized fear of snakes.

Given Wilson's status as a scientist, naturalist, and fellow snake-lover, you can imagine my excitement in meeting him, particularly as I was a new faculty member at Duke University still trying to get my bearings. The day I was to have lunch with Wilson, along with other faculty, I decided to swap my typical field clothes for a dress, as a way of showing respect. I hated dressing up, and my face was puffy and painful from postpartum rosacea; I was feeling particularly self-conscious.

When it was time to meet Wilson, I walked into a classroom, grabbed a plate, piled it up with salad, and sat down at a long rectangular table, right next to Wilson. I had gotten there early so only a handful of other folks were in the room. Wilson had gentle eyes, and I had questions outlined for him in advance, which I started to ask soon after sitting: *What skills do you think naturalists need to develop? What is the role of the naturalist in science today? What advice would you give to an aspiring ecologist?* The basics. Wilson answered the questions patiently and thoroughly, and eventually it was time for him to chat with another young, enthusiastic faculty member. I thanked Wilson for his time, and then he turned to the woman on his other side, about my own age, and said to her, "Now you have the look of a healthy field ecologist."

Wilson's words constricted my heart and my sense of self. The truth was *I* was the field ecologist, and the woman he had turned to worked on computational ecology; she didn't work in the field at all. But she was dressed casually and her skin wasn't puffy and she did look healthy. I was stung. I had talked to one of my idols and failed to convey my core values; I had talked to one of my idols and failed to show him what we had in common; and honestly, I had talked to one of my idols and failed to impress him.

As Brené Brown would say, this was a facedown moment, and I would have to grapple with the extent to which I judged my self-worth based on the feedback of others. Unfortunately, I wasn't ready to grapple with that then. Nope, I've had additional awkward encounters with near-idols. I once started a correspondence with a humanities scholar in the hope of finding a mentor. The correspondence quickly devolved into tawdry flirtation. I felt frustrated and bitter; the relationship disintegrated along with my idolization. It is very hard, when we get right up close to a figure we admire, not to see their warts and wrinkles along with our own. It can happen with dead idols too.

As an eleven-year-old, I became passionately interested in Thomas Jefferson. I had visited Monticello on a family trip and was struck dumb by the magnificence of the Palladian-influenced home, the library filled with Herodotus and Humboldt, Shakespeare and Sterne, the whimsy of an alcove bed, the light streaming into a room filled with busts and paintings, the naturalistic garden once alive with *Alamode* double-blue hyacinths and *Baguet Rigaut* tulips, the beds of heirloom vegetables, and the towering Tulip Trees. I was fascinated by a man who could change the course of history with his provocative words and powerful vision and still find time to cultivate so many interests and imbibe so much knowledge.

Oh, I fell hard for Thomas Jefferson. So much so that rather than having posters of New Kids on the Block or Kurt Cobain on my bedroom wall, I had a poster of Thomas Jefferson, his white hair curling slightly on his forehead, his eyes steady and challenging, his jaw strong and tense, a white cravat snug around his neck, underlying a black coat.

Jefferson himself was a nature lover. Throughout his life, he relaxed by taking long walks in nature or riding his horse across Monticello. He was the first Anglo-American to try to list all of Virginia's birds. He was the first to write a scientific description of the Pecan tree (*Carya illinoiensis*). Even as president, while taking meteorological data, Jefferson made basic phenological annotations: "weep[ing] willow leafing" and "dogwood blossom[e]d" and "frogs sing."

It's unclear what Jefferson thought about snakes. In 1822, Jefferson used the kite, a bird of prey, and the snake as a metaphor for the war between Russia and Turkey, as he explained to John Adams that "whichever destroys the other, leaves a destroyer the less for the world." And while Jefferson himself didn't contribute to the wealth of Western scientific knowledge about snakes, he launched an enterprise that did: the Lewis and Clark expedition. The expedition had a number of encounters with rattlesnakes in particular, seventeen alone in Montana, which usually ended badly for the rattlesnake. Expedition member Joseph Fields seems to have been bitten by a rattlesnake, and Meriwether Lewis is credited with having contributed the Prairie Rattlesnake, *Crotalus viridis*, to the annals of Anglo-American science. In his records, Lewis provided a count of the snake's ventral and subcaudal scales, important ways of distinguishing among species; his collection was eventually described by Constantine Rafinesque back east in Philadelphia.

There is much to admire in Jefferson, but as my own interests broadened and my reading deepened, I began to see the warts and wrinkles and worse. As historian and author Joseph Ellis describes in *American Sphinx,* Jefferson was an opponent of strong executive power who wielded that power to double the size of America, a visionary who failed to self-reflect, and most disturbingly, a voice of liberty who enslaved others. Beginning to see Jefferson for who he was, and Monticello as the forced labor camp that it was, also meant that I needed to see myself for who I was, a young white woman who had the privilege of following her passion for nature and snakes, and the privilege of enjoying an idealized version of Jefferson for many years before coming to grips with his cruelty, contradictions, and the way those contradictions are currently reflected in today's white supremacy culture.

Another idol I clung to, and cling to still, is Robert Kennicott. Today, his bones lie on display on a bed of quartz fragments in a glass case under the yellow lights of the Smithsonian in Washington, DC. His skull shows off strong, straight teeth and a well-shaped head, once covered by rather long, thick black hair; it also reveals a life lost too young.

Robert Kennicott was born in 1835 in New Orleans, Louisiana. His parents were both New Englanders, and his father's work as a doctor had brought the family to New Orleans to combat a yellow fever epidemic.

The family would move again. This time up the mighty Mississippi. In 1836, they arrived in northern Illinois, less than ten miles from the shores of Lake Michigan and twenty miles northwest of Chicago, a burgeoning city that would be incorporated the next March with a population of about four thousand people. For twenty years, the growing family lived in an ever-expanding log cabin in a Bur Oak and Shagbark Hickory grove surrounded by a sea of endless tallgrass prairie. They called it "The Grove," and it is known as such today, preserved as a National Historic Landmark.

Robert was a natural naturalist, exploring several hundred acres of prairie, marsh, and wood in search of all manners of creatures, including fish, insects, amphibians, and mammals, despite often being sick as a child. By age eighteen, Robert had studied with Dr. Jared Kirkland, a renowned ornithologist in Ohio; had begun sending specimens to the Smithsonian Museum; and had conducted experiments on the effects of rattlesnake venom, where he showed that the Massasauga rattlesnake's venom was harmless if ingested and that the Massasauga itself wasn't affected by bites from its own kind. Robert kept several Massasaugas in captivity and extracted the venom onto a silver spoon himself. This sort of work helped Robert earn one of his nicknames, "Bob the Serpent Tamer." In fact, Robert had quite a penchant for snakes, going so far as to advocate that the rattlesnake serve as the United States' national symbol instead of the Bald Eagle, which he held in low regard.

For a suburban kid growing up outside Chicago in the 1980s and 1990s, it was unusual to be so taken with a naturalist from the early 1800s. As a young girl, I played with Barbie dolls, always trying to make the play richer. At first, I imagined the dolls on some grand adventure, often in the deserts of the American West. Later, when my family would visit craft fairs around the city, I'd ask my mom to buy Barbie-sized hand-sewn dresses with old fashioned floral prints reminiscent of another time. I also remember being limited to the backyard, and calling two or three neighborhood boys to the fence, begging them to find something interesting and bring it back to me. This usually meant I got a jar full of tent caterpillars. I was always seeking something more. Then I discovered Robert Kennicott, who represented all the things that are missing or scarce in modern metropolitan life: adventure, connection to nature, and quiet.

I could picture Robert clearly on the lawn in front of the Gothic Revival home his father built, watching writhing rattlesnakes in a pit he had dug especially for that purpose. I could feel the tingle up my spine from proximity to a dangerous creature. I could intuit the comradery that develops when watching the rattlers go about their business, fighting, mating, eating, and sleeping; the comradery that develops observing snakes doing what we ourselves do. I could imagine the space as it was, without the honking horns of modern-day Milwaukee Avenue or the constant drone of the tollway or the jet engine noise of planes going back and forth from O'Hare, then one of the world's busiest airports.

There was an essentialism to Robert's education, too, that the average American kid lacks. Robert didn't have to stand at the corner and wait for the school bus. He didn't have to sit on that school bus with the chaos of childhood, cramped and close-quartered, driving for an hour around town. At worst, Robert walked to the new schoolhouse down the road, but mostly he was tutored in the comfort of his own home. Robert's curriculum was simpler too. He wasn't rushing back and forth from one sterile, fluorescent light–lit classroom to the next, switching gears from Health to Biology to Driver's Ed to Statistics. He was likely reading from a recently secularized text, the William Holmes McGuffey's readers.

Perhaps most striking, however, was the intellectual open land that was ahead of Robert. Today, a student interested in natural history wades through volumes of literature in a daunting search for a new area of inquiry, simply to find an aspect of natural history that hasn't yet been described for a particular organism. For Robert, the field of natural history was still wide open. So many big, beautiful vertebrates, macro-scale organisms that you could see with your naked eye and grab with your bare hand, had yet to be described by Western science. And the behavior of hundreds of species had yet to be seen and recorded and described by European Americans making their steady, destructive march over the continent.

When Robert was born, there was no Shreveport, Louisiana, or Houston, Texas. By Robert's death, in 1866, there were still only thirty-six states in the Union. Nebraska and Colorado and the Dakotas had not yet been added to the docket. In fact, this sense of expansion and exploration—the same sense that undergirded so much of the harm done by European colonization of the New World—would fuel Robert's most well-known work: the exploration of the Alaska territory, then known as Russian America.

Robert was only twenty-nine years old when he was selected as a scientist for the Western Union Telegraph Expedition, an enterprise primarily meant to find a telegraph line route linking North America and Russia, one that would stretch over the Bering Sea. The telegraph was never constructed, as the future lay with undersea communications cables. But the expedition was a success in other ways, at least as defined by the mores of colonialism. Robert collected artifacts from Native Peoples, including a doll, pipes, and a finely woven, watertight basket that are still housed today in the Smithsonian, and he contributed valuable information on the geography and natural resources of Alaska, which helped Senator Charles Sumner convince the US Senate that the Alaska territory was worth buying from the Russians in 1867.

The expedition, however, came at a great cost. At only thirty years old, Robert was found dead on the beach of the Yukon River, his pocket compass lying beside him with lines drawn in the sand showing the bearings of the mountains that lay ahead. Robert Kennicott's death was a mystery that haunted researchers and aficionados for nearly 150 years. Did he commit suicide? Did he die of natural causes? In 2001, Smithsonian scientists traveled back to The Grove and opened Robert's cast-iron coffin. Eventually, they took his remains back to Washington, DC, for tests. Robert hadn't committed suicide at all; he was the victim of a weak heart under intense physical stress.

Alaska would become a state more than ninety years after Robert's death. For Robert Kennicott's contributions to its founding, Alaska's colonizers would memorialize him with the Kennicott Glacier, Kennicott Valley, and the Kennicott River. Western science would memorialize his work with the scientific names of the Stripetail Darter (*Etheostoma kennicotti*), Broad Whitefish (*Coregonus kennicotti*), Kennicott's Neptune (a mollusk, *Beringius kennicottii*), Western Screech Owl (*Megascops kennicottii*), Alaskan Arctic Warbler (*Phylloscopus borealis kennicottii*), and more.

As a thirteen-year-old, I began to volunteer at The Grove—the moniker forevermore associated with Robert's prairie home. I spent three years volunteering in the nature center that now sits on the property, and I spent four more years employed as a part-time naturalist and historic interpreter of Robert's 1856 house, one that Robert's parents finally built after twenty years in a log cabin with seven kids.

Robert's home was built by his father, Dr. Jonathan Kennicott, whose own life was an adventure. When Dr. Jon moved the Kennicott clan to establish themselves in rural Illinois, he served as a doctor, horticulturalist, and the editor of the *Prairie Farmer* magazine. His Gothic Revival house stood as a testament to his wide-ranging interests. The brown board-and-batten edifice, with its steeply pitched gable roof and five sash windows framing the front door, offered stunning views of his unusual flowers and fruit trees, plus a grove of oaks and hickories and a prairie that sloped down to the Des Plaines River.

The crowning glory of the house sits atop the portico that protects the front door, a small, dormered room that was Robert's lab. As a young volunteer, I'd stand in that cramped room and stare at the artifacts representing Robert's life: jars of formalin-bound snakes and frogs, artifacts from the Indigenous people whom Robert once knew, and even a pair of wooden snowshoes. The artifacts whispered to me: *there's more out there for you to see . . . go see it.* I imagined myself in Alaska, tromping in the snow on unwieldly snowshoes or climbing a morainal hill in northern Illinois, a moist woodland covering the sandy rise, my hand reaching out to grab a delicately flecked Blue-spotted Salamander.

The single window of Robert's lab also provided a perfect vantage point from which to overlook the stuff of a herpetologist's dreams—that snake pit in the front yard, once writhing with Massasaugas, whom Robert called "queer pets, but by no means uninteresting ones."

Known as the "swamp rattler" or "black snapper," the Massasauga is a fairly diminutive pit viper, reaching adult lengths of only two and a half feet. With dark-brown blotches on a background the color of dried wheat, it boasts few other adornments, just a few rows of smaller spots down its side and a black-and-gray mottled belly.

It's hard to know if Massasaugas are really the quintessential rattlers of the tallgrass prairie peninsula. Overlays of the pre-European extent of the tallgrass prairie only match up with about a third of their current distribution. There's so little prairie left today that their current habitat—the edges of forests and grasslands, and in sedge meadows and wet prairies—might just be the closest they can get to wide-open fields of Yellow Prairie Grass (a common name based on the translation of a Lakota term for *Sorghastrum nutans;* the mainstream English-language common name is culturally insensitive) and Big Blue Stem. Robert Kennicott himself called them "prairie massasaugas" and claimed they could be found living in the holes of prairie dogs.

Regardless of what their preferred habitat *was,* it is certainly gone now; only six to eight relict populations of Eastern Massasaugas remain, stretching from Iowa to central New York. While researchers have been noting the precipitous decline of Massasauga populations for the last ninety years, the species is only listed as a candidate under the United States Endangered Species Act.

Between human malice and hogs, roads and urbanization, and even mechanized agriculture, Massasaugas continue to be at risk. Plus, new threats have shown up on the landscape: OO, or *Ophidiomyces ophiodiicola.* OO is an opportunistic fungus, also known as Snake Fungal Disease, or SDF. Like other fungal pathogens affecting herpetofauna (think: Chytrid fungus and frogs), the threat seems to be broad, affecting pit vipers, colubrids like ratsnakes, and other reptiles. The disease outbreak appears to be especially detrimental to the already struggling Massasauga.

The way OO causes mortality in snakes is gruesome, with one report describing snakes affected with disease as mummies where "tattered skin [clings] to the animal's warped face." The fungal infection makes the skin rough and bumpy, constantly sloughing off. OO can also cause myositis, which inflames and stiffens a snake's muscles, and osteomyelitis, a bone infection. Over time, the disfigurement caused by the lesions, myositis, and osteomyelitis makes it impossible for snakes to hunt or swallow prey. Often when people come across a snake with OO, they first notice that it is emaciated, slowly starving to death; sometimes they notice that the snake isn't breathing well. That's because OO also causes pneumonia. The extent to which OO will cause declines in already suffering Massasauga populations is unknown, but fungal pathogens have caused extinctions in other rare species (for example, the Sharp-snouted Day Frog, *Taudactylus acutirostris*), and researchers are worried.

While it might be hard to rally support around the Massasauga, it being a rattlesnake and all, no one has ever reportedly died from its bite. Robert Kennicott recognized that intolerance of rattlers was largely unfounded, saying: "Were it not for foolish prejudices, we might see much to admire in the rattlesnake. The rattlesnake was for a time our national emblem, and it is to be regretted that it was so soon thrown aside for the bald eagle. For despite the horror in which he is held, the reptile is by far the nobler animal of the two. He is no impotent and cowardly robber, like our emblematic bird—makes no unprovoked attack,—and always sounds his warning rattle."

If Robert's snake pit and unapologetic defense of rattlesnakes wasn't enough to catalyze idolatry from a twentieth-century teenager, this apocryphal story I learned as a guide in the Kennicott house sure was: Legend had it that one day Robert's family was sitting around their long, wooden dining room table with a couple of neighbors. The family was waiting for Robert before eating their supper, the mother in her plaid daydress with its wide, crocheted collar, the father, a doctor, black-haired and handsome still, and sisters Alice and Cora with dark ringlets and piercing eyes. Suddenly, Robert bursts in. He declares that he is famished, and he sets down his day's work on the dining room table: a live Massasauga rattlesnake.

While we're on the subject of rattlesnakes, biologist David Steen neatly explains the possible origin of their rattles. Rattling likely developed from tail shaking, because many snake species shake their tails. Some snakes may have had a little extra tissue on their tails, which made more noise when they shook them, thus scaring off more predators. Since these snakes survived, they also reproduced more. Over time, differential survival and reproduction selected for more tail shaking (rattling) and more tissue (rattles). I'm sure the folks at Kennicott's dining room table would have loved that fun fact.

It may not be surprising that my two idols, Robert Kennicott and Donald Culross Peattie, are related. Robert Kennicott grew up at The Grove with six siblings, including those two remarkable sisters with dark ringlets, Alice and Cora, who helped him assemble his natural history collection. Alice was a better shot than her brother and was handy for collecting bird specimens, at least one of which was sent off to the Smithsonian for zoological description. Cora also contributed her services to Robert, offering to catch bugs and snakes for her older brother. When Robert died, she carried a newspaper article reporting his death with her for the rest of her life.

Besides being a remarkable horsewoman and enjoying the study of algebra, Cora serves as an important link in our story. It was Cora who married James Redfield, the family tutor who left to finish his law studies and came back to wed his "angel in pantelettes" in the middle of the Civil War. And it was Cora who gave birth to three daughters and one son, Robert Redfield, who became a lawyer like his father. Robert Redfield and Bertha Dreier, daughter of the Danish consul, had two children. One of these children, Louise Redfield, was an author in her own right and the wife of Donald Culross Peattie. This makes Robert Kennicott the great-uncle-in-law of Peattie, and Peattie was well aware of this important connection. Peattie devoted an entire book to the Kennicotts and The Grove.

While Peattie respected the story of the Kennicotts and The Grove, it didn't mean that he shared all of Robert's predilections. In fact, Peattie himself described an encounter that he had with a rattlesnake at a ranch in the Mojave Desert. After sunset, on his last evening at the ranch, Peattie took a walk and came across a rattlesnake. The snake was calm and watchful, not reared back or ready to strike. Peattie debated with himself. On the one hand, he was inclined to let the snake be, and on the other, he felt he should protect the kids and dogs at the ranch. Peattie trudged back to the ranch house to grab a hoe. He walked back with the gardening tool, and the rattlesnake lay unmoved. Then the snake lifted his head, but did not strike; instead, it fled into the bushes. It began to rattle, agitated and scared. Peattie stretched his hoe into the bush and "hacking about, soon dragged [the snake] out of it with his back broken." By the time I read this about Peattie, I had already learned my lesson—seek not perfection in others. We may idolize traits and deeds, but never men.

Snakes aren't hard to kill, and that has much to do with their basic anatomy. Even relative to other reptiles, snakes are known for their fragility. Their heads are delicate, owing to a structure that allows for maximum flexibility of the mandibles so snakes can eat prey larger than their heads. And their spines include eighty to four hundred vertebrae, and after the first two, each with a delicate pair of ribs. The vertebral column and the ribs protect elongated organs, including a long lung and stretched-out kidneys that conform to the snake's body. In some arboreal species, which are notoriously long and attenuated, even the heart is elongated. While this elongation provides remarkable variety in the ways that snakes can move and the environments they can move in, it also leaves them vulnerable. A cracked rib can easily pierce a vital organ, and with such a long vertebral column, it's relatively easy to hit a snake and break its back. Peattie knew this too. In a book review that Peattie wrote of Ditmars and Bridges's *Snake-Hunters' Holiday,* he shared that snakes' "backbones are frangible as thin porcelain."

Snakes are easy prey, for humans at least, as rattlesnake roundups attest. From Pennsylvania to New Mexico, January to July, rattlesnake roundups occur across the United States. Most are organized by local Chambers of Commerce, Lions Clubs, or Jaycees. The events often have cutesy names like the Opp Rattlesnake Rodeo or the San Patricio Rattlesnake Races or the Mangum Rattlesnake Derby. But make no mistake, these events are anything but cute. They can have devastating effects on rattlesnake populations.

In the mid-1990s, researchers identified at least seventeen "rattlesnake roundup communities" within the range of the Western Diamondback Rattlesnake (*Crotalus atrox*) in Texas alone. The roundup in Cleburne, Texas, was so "successful" that it has been canceled because they "could not find enough snakes to collect." The roundup in Sweetwater, Texas, is notorious. By the second weekend of March each year, local hotels in Sweetwater are sold out to accommodate more than thirty thousand tourists who come for what is billed as "the World's Largest Rattlesnake Roundup." Over the last sixty years, this roundup has hauled in more than 132 tons of rattlesnakes. In 1982, nearly eighteen thousand pounds of rattlesnakes were killed during the four-day event. In 2019, by midmorning of the opening day, they had already collected four thousand pounds.

The Sweetwater Rattlesnake Roundup hauls in more than just snakes; it brings in a lot of money: about $8.3 million. An all-day pass costs thirty dollars per person, and it costs sixty dollars to enjoy the entire weekend, filled with events like the "Miss Snake Charmer Pageant," "Guided Hunt," and the "Snake Eating Contest," where snakes are "killed with a machete and skinned and gutted in front of the crowd and the meat is then deep-fried." Even during the 2020 coronavirus pandemic, the Sweetwater Rattlesnake Roundup continued. In March of that year, children peered through Plexiglas windows to see gaitered men standing in a tank of more than one hundred rattlesnakes or watch a Sweetwater Jaycee milking venom from one of those snakes. Others gaped as a line of children and adults alike skinned rattlesnakes tied to ropes and hung from poles.

While Sweetwater Jaycees say they hardly make a dent in the rattlesnake population, a study from the eastern United States suggests that roundups have depleted local Eastern Diamondback Rattlesnake populations, forcing hunters to travel farther to collect them and resulting in a decline in both the numbers and weights of the largest snakes turned in over the last two decades. J. P. Jones, founder of a rattlesnake roundup in Opp, Alabama, is quoted saying: "We have to drive a hundred miles from here. . . . They [are] a lot scarcer now. When I started [in 1959] we just hunted in the woods here. We used to get fifteen a day. You won't get none today."

Eastern Diamondback Rattlesnakes historically occupied Longleaf Pine ecosystems, the extent of which has been reduced by about 97 percent in the southeastern United States. Habitat loss has already pushed the northernmost boundary of the Eastern Diamondbacks' range southward, and the species is threatened or rare in North Carolina, South Carolina, Alabama, and Mississippi, and possibly extirpated entirely from Louisiana. Expanding agriculture and increased traffic have likely contributed to declines. Rattlesnake roundups don't help either.

Research shows that the effects of rattlesnake roundups on snake populations goes beyond simple mortality. Roundups perpetuate negative stereotypes about snakes and normalize their killing. Some have argued that these types of experiences send disturbing messages to children, citing the link between childhood violence toward animals and interpersonal adult violence.

But there is good news: roundups can be reimagined. A roundup in Fitzgerald, Georgia, has become a Wild Chicken festival that has "experienced enormous success," and one in Claxton, Georgia, now celebrates wildlife instead. Others, like in San Antonio, Florida, have disappeared completely.

Perhaps there is hope. Even those who have been trained from birth to believe that "the only good snake is a dead snake" have experienced guilt and remorse. D. H. Lawrence, in what one can only assume to be an autobiographical poem, wrote of a snake he saw one day in Sicily, while he was still in his pajamas, which came up to a water trough. The snake "rested his throat upon the stone bottom, / and where the water had dripped from the tap, in a small clearness, / He sipped with his straight mouth." Lawrence noted that the voice of his education said, "he must be killed" but that deep down he liked the snake. Even still, all those inner voices came back to him, calling him a coward if he didn't kill the snake. He picked up a log and threw it at the water trough. The snake convulsed and writhed but slid away. Lawrence did what his culture had taught him to do but then noted: "And immediately I regretted it. / I thought how paltry, how vulgar, what a mean act! / I despised myself and the voices of my accursed human education."

The hope lies in humankind's capacity for empathy and change. The hope lies in our ability to change the nature of our "accursed human education" and create cultural conditions of compassion and empathy for all creatures. And perhaps we can work together to change the current culture and, as Lawrence says, allow the "king in exile," the snake, "to be crowned again."

LOVE AND LOATHING

AT AGE FIFTEEN, I had volunteered for more than two years at the one-hundred-plus-acre nature preserve north of Chicago. I passed much of my time in the interpretive nature center, a modern, two-story 2,500-square-foot log cabin. I spent hours in this building, watching chubby Tiger Salamanders greedily eating crickets or scrubbing the glass front of an animal tank free of the little fingerprints left by children the previous day.

On Saturday mornings, I'd walk into a long, narrow workroom that stretched along the backside of massive tanks. Three of those floor-to-ceiling tanks held fish: long-snouted Gar, glassy-eyed Crappie, and broad-headed Bullheads. The last tank was different, split into two parts, top and bottom. The top was home to a shy, red-phased screech owl, and the bottom held Massasauga rattlesnakes, those weak-venomed prairie denizens. I always greeted them first.

The workroom smelled like dried fish and musty, male mice and pungent, earthy crickets. There were hundreds of animals to feed, and I was inhaling the varied aromas of their food. An old refrigerator stood near the primary entrance to the workroom, and a faded magnet held the list of daily tasks to the door. Saturday was a "minnow day," one of three days of the week that we scooped heavy nets full of minnows into the fish and watersnake tanks.

Minnow days were exciting, but hard work. I took two five-gallon buckets down into the dark basement, where the filtration system for all those tanks upstairs was housed. The minnows lived down there too, in a green tank at hip level with an open top. It was a couple feet wide, four feet long, and as deep as my entire arm. The icy water chilled the bones in my hand. Sometimes, a fellow volunteer and I would see who could hold their hand in the longest, and I still remember that awful cold and the frustration of losing week after week.

On minnow days, I dipped the buckets into the tank of silvery fish, filling them halfway with that cold water. Then I scooped at least six nets full of small fry into each bucket along with one large sucker, a fish twenty times the size of the others. Writhing in the buckets, the minnows were dark and dense. If I brought my head down close, I was splashed in the face by the wild movements of all those wriggling fish.

The next step was to carry those buckets, one in each hand, back upstairs. When I first started volunteering, I had to stop halfway on the landing to set down the heavy loads and rest my weak arms, but two years in, I could plod up those stairs, buckets full of water and fish, and not stop until I was back at the workroom.

Once there, I'd grab a step stool and drag it in front of each tank, dumping in a couple half nets of minnows for big river fish, the green-hued Large-mouthed Bass with their gaping mouths and the pouting Flathead Catfish with barbels sticking out of their chin forming a straggly beard. The nature center had other tanks too, one full of hungry Lesser Sirens, another with an Alligator Gar and an Alligator Snapping Turtle. That big sucker was for the Alligator Snapping Turtle, a one-hundred-year-old behemoth from the Mississippi River that we fed with long tongs.

On minnow days, I saved the best for last: the watersnakes. The watersnake enclosure was made entirely of Plexiglas and stretched from floor to ceiling, six feet across. The bottom half was a fish tank, filled with sinuous snakes and a few minnows that had so far evaded their predators. The top half had a platform that propped up a tangle of branches on which the watersnakes could bask closer to the three warm lights near the top.

This bright enclosure opened by sliding a piece of the Plexiglas across the front of the tank. Sometimes the snakes would sit on the platform with their heads lined up at the corner where the sliding door would first open up, having recognized the thrashing of the minnows in the bucket or the bright-green net. It was precarious, simultaneously holding the net, pushing back the eager snakes lest they escape, avoiding being bitten, and sliding the reluctant glass.

I eased open the glass just enough to dump in one scoop of roiling minnows. They splashed into the water in silver streaks that lured the quick watersnakes. The watersnakes would slide into the water, plunging deep to secure a minnow. They would grab with an open maw, swallow the minnow headfirst, and soon the silvered flash was absorbed into the snake. Bodies merged. One life had become another. As Andreas Weber suggests, consuming another being becomes an intimate show of interdependence.

The snakes cut through the water in search of prey; the minnows moved in a frenzy, directionless and lost. I would watch until the water's surface no longer frothed with activity, and then I'd return my bucket and net to the workroom.

By now, it was time for a break, and I'd move under the grand wooden beams of the interpretive center, basking for a moment in the filtered shafts of sunlight that came in through the high windows, finding my seat on a wooden bench directly in front of that exquisite snake tank.

I fell in love with the physical form of those snakes. At first, I was rapt over the array of colors present in just a few species: crimson bellies, copper backs, brown bodies, and black bands. Then I noticed the patterns: a single swathe of pure color, thick stripes at regular intervals, like the shadows cast on the ground in front of a picket fence, or a complex patchwork of squared splotches connected by slender, jagged lines, making delicate diamonds outlined in black, extending from the snake's plain, tan neck to the tip of its pointed tail.

Then I watched how their bodies contracted in tension, released at rest; how minute movements propelled them down into the water or lithely across its surface; how simple pulses pushed them forward along a thin branch and let them balance perfectly in the light.

Sometimes I would bring a notebook and write. In August, when I was fifteen, I overflowed with admiration and envy for these creatures, writing:

> As I watch them, I see the qualities they have that I wish for me. Graceful and elegant with every contraction of their powerful muscles. They glide through the water with a superior air, they don't make a sound, but their presence is felt. Powerful and having the ability to bring human fear to the surface with one simple move to strike. They resemble well-respected royalty walking through a room with their heads held high and movement sculpted by grace. Yellow-bellied Watersnakes: [Evoking] ancient Greek statues; their form is simple, lines fluid, but their façade is stoic. This simplicity is powerful.

I wasn't the only one inspired by the graceful lines of snakes. Works of art from around the world abound with images of these creatures, redolent with symbolism and beauty. During the Edo period, Japanese artists carved wooden icons with the coiled body of a snake and the face of an old man, representing Ugajin, a deity of fertility and bounty. In eighteenth-century Burgundy, a French silversmith forged a *taste-vin*—a shallow cup used to taste wine drawn from the barrel. Its handle was in the form of a miniature snake, giving a nod to Bacchus, the god of wine, and Hygieia, the goddess of health.

Today, snakes remain powerful in art. The Dominican American artist Firelei Baez forms a beautiful Gordian knot of lacey, white snakes. This snake knot represents a tignon, a turban-like head covering, an image that Baez overlaid onto graph paper from the American Sugar Refinery in a powerful oil and acrylic piece. Even Banksy, the mysterious British street artist, has incorporated snakes into his work, using fiberglass, resin, and acrylic to make a python-patterned sculpture that seems to show a snake after having consumed a particularly big meal, leaving a Mickey Mouse–shaped lump in the snake's body.

Snake symbolism allows us to grapple with the human condition, and in my own adolescent notes, I did the same. I remarked on the snake's proud nudity, not understanding the shame associated with nakedness in our human society. I struggled with contemporary conceptions of beauty, longing for widespread recognition of the simple and natural. And I began a deep dive into the underworld of gender roles and those subtle signals of my place in it; a shadowy realm absent among my beautiful watersnakes.

Mostly, though, I began to think about fear.

Once, while I sat on that hard wooden bench with an amazing view, a woman was entering the nature center. She was tall and big-boned but struggled to pull open that heavy wooden door. She stepped over the threshold, her gaze rose over the low, fossil-lined turtle enclosure, past the flighty kestrels, and landed on my beloved watersnakes. She screamed and then panically repeated, "oh no, oh no," as she backed out of the door. The woman was so terrified of snakes that she refused to reenter the building. I was perplexed by her reaction and felt rejected by the disgust she showed for the snakes I loved.

Yet, research suggests that this fear of snakes isn't uncommon. In fact, some studies suggest that fear of snakes and spiders is innate, harkening back to our days in the trees as petite primates. Children as young as five months old have different reactions to images of snakes and spiders than to those of, say, flowers. Fear of these animals elicit more severe phobias in a relatively large segment of the population: a full 5.5 percent of people have snake phobias.

Psychologists and researchers Arne Öhman and Susan Mineka may have summarized human fear of snakes best:

Snakes are commonly regarded as slimy, slithering creatures worthy of fear and disgust. If one were to believe the Book of Genesis, humans' dislike for snakes resulted from a divine intervention: To avenge the snake's luring of Eve to taste the fruit of knowledge, God instituted eternal enmity between their descendants. Alternatively, the human dislike of snakes and the common appearances of reptiles as the embodiment of evil in myths and art might reflect an evolutionary heritage. Indeed, Sagan (1977) speculated that human fear of snakes and other reptiles may be a distant effect of the conditions under which early mammals evolved. In the world they inhabited, the animal kingdom was dominated by awesome reptiles, the dinosaurs, and so a prerequisite for early mammals to deliver genes to future generations was to avoid getting caught in the fangs of Tyrannosaurus rex and its relatives. Thus, fear and respect for reptiles is a likely core mammalian heritage. From this perspective, snakes and other reptiles may continue to have a special psychological significance even for humans, and considerable evidence suggests this is indeed true. Furthermore, the pattern of findings appears consistent with the evolutionary premise.

Moreover, women have been documented as being four times more likely than men to have fears and phobias of snakes and spiders, but not of other things that commonly scare people, like heights or injections. Hypotheses abound about why women and men might differ in their incidence of snake fear. Some researchers have proposed that the evolutionary or fitness costs of being bitten by snakes would be greater for women because a mother's death from snakebite would likely result in her children's deaths from lack of care. Another hypothesis is that the cultural transmission of fear is promoted more among women than men. Perhaps both mechanisms are at play: compared to boys, eleven-month-old girls more quickly learn the relationship between negative facial expressions and snakes.

Fear of snakes might have more consequences than we would initially suspect. Fear of snakes among teachers-in-training negatively influences their conservation attitudes and decreases the likelihood that they will incorporate snakes into their future curricula. This is a shame because research also indicates that when people interact with snakes, they are less likely to think of snakes as threats and more likely to have positive attitudes toward them.

For a psychologist, the immediate fearful reaction of the woman who entered the nature center wasn't completely unexpected. For me, this woman's reaction was strange and disturbing. I knew that sometimes the public was hesitant to touch the snakes that I held out for them, but I had never seen such an automatic or intense manifestation of that fear.

I began my campaign against fear that day. It wasn't that I sought to eradicate fear itself. No, as a high school student submerged in a sea of adolescent fears—friends afraid to talk to their crush or wear what was simply comfortable or express their creative side—I recognized that fear itself had a solid foothold in the human ecosystem, as entrenched in our ecology as the Brown Tree Snake has become on Guam. The fear itself, like the Brown Tree Snake, might be impossible to completely eradicate, but the effect of this invasive element could be ameliorated.

The Brown Tree Snake (*Boiga irregularis*) is strikingly beautiful. Its base color, that is, the color of its back, is like baked cinnamon biscuits, with delicate black lines inked across its neck and upper back. Its eyes, too, are haunting and catlike, with elliptical pupils set in honeyed spheres.

Originally found in many of the nations of Oceania, including Australia, Indonesia, and Melanesia, this rear-fanged, climbing snake actually didn't go far when it arrived in Guam, a 210-square-mile U.S. territory that is also the largest island in Micronesia. The problem was, until sometime in the late 1940s or early 1950s, this island had never seen snakes before. Now, there are at least two species of snake in Guam, the Brown Tree Snake and the Brahminy Blind Snake, which impacts native termites and ants. They have few predators. The lack of predators allowed numbers to quickly swell to two million Brown Tree Snakes, up to five thousand snakes per square kilometer.

Moreover, the birds and bats of Guam hadn't evolved defenses against snakes. This means that since the U.S. military vessels accidently brought the Brown Tree Snake to the island, at least twelve species of birds and the Little Mariana Fruit Bat have gone extinct in only a few decades. Some of these species were found nowhere else on earth, including the Guam Flycatcher and the Ko'ko', or Guam Rail. The Guam Flycatcher, a beautiful blue-black bird with a white-and-buff breast, was always secretive, hawking insects in dark lime-stone and ravine forests. Now the bird is lost to the world, last seen in 1983. The Guam Rail, by contrast, was flightless, its brown-ochre plumage and barred breast feathers camouflaging it in scrubby, secondary growth forests. As a ground nester, it was especially vulnerable to predators, and it disappeared from Guam by the late 1980s. Luckily, the species still exists in captivity, being bred at American zoos and by the Division of Aquatic and Wildlife Resources on Guam. The story isn't much different for the Little Mariana Fruit Bat, or Guam Flying Fox, with an impressive wingspan of over two feet. This species has been extinct since the 1970s.

The decline and extinction of bats on Guam has had cascading ecological effects. Researchers have found that tree recruitment, or the growth of seedlings, has declined by as much as 92 percent. Seeds—once carried far afield by fruit bats—now aren't dispersing beyond the shade of their parents. In fact, 66 percent of Guam's trees relied on animals to distribute their seeds.

The thing about Brown Tree Snakes is that they endure. Using the ecology of its native range as a guide, researchers have identified at least a few predators that could control Brown Tree Snake numbers, including the Red-bellied Black Snake and Cane Toad. The problem is, introducing these two species to Guam might have even more devastating consequences. The Cane Toad was introduced to Australia to control the Cane Beetle and French's Beetle, which were injuring sugarcane crops, but the introduction of Cane Toads to Australia did nothing to control the beetles, and the toads themselves are associated with declines in mammal, goanna, and snake populations.

In Guam, the U.S. Department of Agriculture has spent more than $8 million on chemical weapons, including paracetamol, better known at acetaminophen. You might be imagining the USDA lacing mice with Tylenol, and that isn't far from the truth. Researchers literally stick 80 mg acetaminophen tablets to the bellies of dead mice. Then, using contraptions resembling modified egg cartons to hold them, the acetaminophined mice are dropped into the forests, and, as one reporter noted, "the snakes do take the bait."

After nearly fifteen years of using a combination of traps, toxicants, and hand-removal to control Brown Tree Snakes, finding a cost-effective solution to the Brown Tree Snake problem on Guam has proven difficult. Only in 2016 were researchers able to complete the first major evaluation of helicopter-dropped Tylenol-mice on Guam's Brown Tree Snakes, documenting reduced snake activity for months after the toxic-bait drop. This method of snake control is catching on, with researchers evaluating the use of baited mice to control California Kingsnakes that were introduced to the Canary Islands.

Snakes can be persistent, and so can fear, but we have myriad ways to deal with both, some more caring and tolerant than others. To manage fear of snakes, we must learn to live with snakes, and perhaps even honor snakes—like the Japanese carvers in the Edo period or the Burgundian wine makers of old. We must expand the sphere of our humanity to the nonhuman as well.

LA SUERTE

 STEPPING OFF THE MINI-coach bus was like stepping into Oz. The world was Technicolor, too vivid for real life. Toucans sang hello with thick bills painted florescent green and orange, tips stained red. Poison dart frogs, glittering like mosaics of emeralds and obsidian, hopped across the wet path. In a daze like Dorothy, I wandered slowly, arriving at a cabin that would be my home for the next couple weeks. The cabin, a slight, wooden construction with a low roof, was raised up on squat stilts. No witches beneath this one. I stepped up onto the porch, but I bypassed the door, drawn to the thick forest behind the cabin. I stared out into the jungle, and a small troop of White-faced Capuchins stared back. La Suerte Biological Research Station, in Limón Province, Costa Rica, was a Munchkinland for students and researchers.

Biological splendor put me to bed that first night—thousands of beetles swarmed the windows, inside and out. Mosquito netting became a psychological refuge from that splendor. At night, the beetles would tap-tap on the solid wood floor, at first masking the slow drops of rain pinging against the corrugated metal roof. The rain would pick up, the roar so deafening that I couldn't hear the person shouting from the bunk across the room. Thunder shook the cabin, kettle drums unleashed until it was nearly dawn.

It was constantly raining, either a light drizzle that managed to get me thoroughly soaked throughout the course of the day or a constant heavy downpour that more efficiently drenched me in minutes. It was still raining the next morning when I trekked three hours through dripping, verdant forest. The charcoal-gray sky endowed the enchanted jungle with a dark, cavernous feel. I heard the big drops plopping loudly as they hit the broad green leaves, before those same drops fell on me, sneaking under my collar, tracing tiny rivulets down my back. I wore a raincoat, but I was soaked to the bone. I wore galoshes, but they were filled to the brim with water.

I walked among the trees, my mosquito-bitten arms brushing against the broad-leaved foliage, my eyes distracted by the drooping vermilion flowers of the Heliconia and the erect brown stilt roots of Neotropical trees. Small Strawberry Poison Dart Frogs hopped across the path. Territorial Jesus lizards clung to lianas interwoven over my head. Proboscis Bats lived in a giant, hollowed old tree. The lowland tropical forests were being quickly converted to banana farms, but La Suerte provided sanctuary for all animals, even snakes. That day, I found a Hognose Viper, curled and almost undetected, under the shrub layer of La Suerte's secondary forest.

The forest rustled and whispered with life during the day, but it downright roared at night. The moist night air held the promise of life—frog calls, staccato pop-pop-pops and trilling harmonies drifted to my ears while strange guttural howls resounded in the distance. I thrived in the darkness, anticipating the next Red-eyed Tree Frog hugging a thin branch or a furry tarantula scurrying from the blinding light of a headlamp. So many frog species were calling that it was difficult to keep up. We found Gray Tree Frogs and Olive Tree Frogs and

Glass Frogs, whose internal organs can be seen through their ghostly thin bellies. Walking down the gravel road in the dark yielded new taxa; I could spy the penetrating red eyeshine of Spectacled Caimans, whose *tapetum lucidum,* an iridescent film behind the retina, reflected the light of my dim flashlight back at me from murky roadside swamps.

If I only spent my days among the Mantled Howler Monkeys and Blue-crowned Motmots, I would have remained fully whole, living into deep contentment and connection with the natural world, delighting in the simple pleasures of watching life thrive. But I didn't. I played soccer barefoot with the locals in town. I sat among college students obsessed with naked photos of moms. I watched a local man carry a black plastic container into camp.

The world is complicated, and snakes aren't particularly loved. In many traditions, snakes are seen only as dangerous or demonic. For example, in the Bible, snakes are frightening, always hiding in the shadows ready to bite a horse or hand. There's the "snake by the roadside, the viper along the path that bites the horse's heels" (Genesis 49:17). There's the warning that "when you demolish an old wall, you could be bitten by a snake" (Ecclesiastes 10:8). Even Paul is bitten, when he "gathered a pile of brushwood and, as he put it on the fire, a viper, driven out by the heat, fastened itself on his hand" (Acts 28:3).

The Bible also uses snakes in moral tales to steer us away from ethical evils, like alcohol and adultery. Wine might sparkle in the cup and go smoothly down, but "in the end it bites like a snake and stings like a viper" (Proverbs 23:32). Snakes also perplex an adulteress, when in a beautiful line, she says, "There are three things too wonderful for me, four that I cannot understand: the way of an eagle in the sky, the way of a snake on a rock, the way of a ship at sea, and the way of a man with a maiden" (Proverbs 30:19).

The snake is also smart, but that intelligence is typically described in a way that undermines that positive attribute. Snakes are "shrewd" and "crafty" and "cunning." "Eve was deceived by the serpent's cunning" (2 Corinthians 11:3). The Bible condemns the snake, explaining that God cursed snakes "above all livestock and wild animals" (Genesis 3:14) and giving people permission to trample them too (Psalms 91:13). People, with the Bible's explicit blessing, have been punishing snakes ever since.

Even when snakes are not presented as evil, things seldom go their way. In the Lowcountry of the southeastern United States—that is, the coastal plain and islands of Florida, Georgia, and South Carolina—live the Gullah, a group of African Americans who have a distinct culture and language intimately linked to their African origins. The Gullah culture arose out of slavery, when the ancestors of today's Gullah-Geechee were enslaved to work on extensive and isolated Lowcountry rice plantations, agricultural forced labor camps. During the American Civil War, while plantation owners were fleeing their properties, the Gullah were organizing to defend their home, and many Gullah served in the U.S. First South Carolina volunteer army, that is, on the Union side. After the war, the Gullah remained relatively isolated in the rural Lowcountry, and their culture continued to develop into the one full of the flavor of okra soup, the tang of herbal medicines, the rich and ringing call-and-response spirituals, and the sweet smell of straw baskets. The Gullah culture is also rich in stories and trickster tales.

In one Gullah tale, a man goes into the woods. He's working hard, chopping wood, but having trouble earning a living to support himself and his wife. A snake comes up to him and, noticing that the man has come on hard times, offers to help with one condition: the man must not tell anyone from where he received help. The man agrees, and the snake literally coughs up money, gold coins spilling from his mouth. The man goes home. His perceptive wife notices something is different. She prods, she presses, and finally her husband tells her what happened. The wife has an idea; she tells her husband to go back to the woods, kill the snake, and cut open its belly, because the snake must have more gold inside its stomach.

The man goes into the woods the next day with his ax. The snake slithers up to him as he's chopping wood, and asks the man how he's doing. The snake is perceptive too; he knows the man has shared the secret. As they're talking, the man raises his ax. He brings it down toward the snake, set on following his wife's advice, but the snake drops off the log quickly. The man misses the snake, and loses control of his ax, accidently cutting off his own leg. In this story, at least the snake ends up unharmed, but snake-human relationships continue to be strained.

That's how it was for the beautiful *Terciopelo* trapped in a black plastic container being marched from town into the La Suerte research station, except the snake hadn't dropped away in the nick of time. The *Terciopelo* sat on a road just outside of a small village. A man saw it. I imagined that he remembered the herpetologists at the research station. He remembered how the older folks in town used to catch animals—sloths, monkeys—and bring them to the station and threaten to kill them. He remembered how the station used to give them money so they wouldn't. The huge *Terciopelo,* clothed in an eye-catching argyle of tan and brown and black, sat on the road. The man beat it with a stick, breaking its spine. He found the black barrel and pushed the dying snake inside. He brought it to the research station. He was angry when the biologist yelled at him. He was angry when no one would pay him. He was angry that he had wasted his time, and I stood wide-eyed, holding my breath to keep myself quiet, trembling with rage over the snake's death and the systems of inequity that had led to it.

Humans have a complicated relationship with snakes. *Terciopelos,* also known as Fer-de-lances, are pit vipers. They live in tropical rainforests and tropical deciduous forests, but they are often found in drier habitats, like thorn forests and savannas, that are adjacent to water. These are places that people tend to live too. While *Terciopelos* often slide away when threatened, they are rather capricious, sometimes choosing to defend themselves rather than disappear. Today, the death rate for bites from *Terciopelos* in Costa Rica is near 0 percent due to effective treatment, but *Terciopelos* are still responsible for nearly half of all snakebites and nearly a third of all snakebite hospitalizations in the country. Understandably, they get a bad rap. The story of *Terciopelos* does have a biblical connection, only the story gets reversed. In South America, *Terciopelos,* making rather strange Davids, are purportedly eaten by Goliath Bird-eating Spiders.

La Suerte was a Garden of Eden. Gardens of Eden for the ten thousand known reptile species on this planet are found all over the world: Central and Western Australia, Southeast Asia, arid Southern Africa, the Arabian Peninsula and North African coast. Even the North American plains makes the list. Snakes disproportionately dominate maps of reptile species richness, in part because their median range is nearly six times larger than, say, that of lizards, even if there are half as many snake species as lizard species on the planet.

When looking for Edens for more than 3,400 of the world's snake species, nearly the whole continent of Australia and parts of Southeast Asia make the cut; so do spots in Southern Africa, the Amazon of South America, and forests of Central America. The hotspots for snake species richness are broadly characterized as pantropical, like the pantropical hotspots of orchids, and, for some, snakes are just as obscenely tempting.

But a pantropical distribution doesn't guarantee safety. In fact, research from the Brazilian Atlantic Rainforest hotspot predicts major range contractions for the majority of snake species in the region. By 2080, more than two-thirds of snake species will lose half, or more, of their original range. These predictions are based solely on changes in climate, that is, changes in temperature, humidity, and rainfall. Being ectothermic, snakes are exquisitely sensitive to these factors, which affect the embryonic development of eggs, habitat and food availability, and a host of physiological processes.

And like all Edens, people live there. More than 145 million people—about 70 percent of Brazil's population—live in the Atlantic Forest area. A population of 1.1 billion people is predicted in Southern Africa by 2050. Australia's population increased by 400,000 in 2018 alone. Some of those people are bad, some good. Some apathetic, some curious. Some cruel, some compassionate. Most people are somewhere in between, and where they fall on these long, tortuous gradients is influenced by a host of factors—culture, psychology, money. But somehow, in nearly every Eden, people are largely ungenerous to snakes. Between the threats of climate change and cultural norms, snakes are in trouble.

OMETEPE DREAMS

I SAT ON A lawn chair, a bottle of dark Flor de Caña rum in my hand, and I sang. I sang a song that was not my own, and would never be my own, but I didn't know that yet. Deep and throaty, the words burst forth over Lake Nicaragua. I sang about the sky, the stars. I was praying or wishing or daydreaming. There was a boy and a dove and a plastic bag. Only the lyrics, thick with desire, and the black depths of rum and the nighttime lake existed.

There was little moonlight that night. The inky lake was hiding its many mysteries, some pouring in from forty different rivers, some trapped when a volcanic eruption closed the lake off from the ocean. Cryptic creatures lurked within the dark waters: Bull Sharks cruising warm, shallow depths for bony fish and stingrays. Swordfish rising silently to the surface at night in search of smaller fish. More than a dozen species of cichlids flashing silver and black, golden-eyed, secretly spawning along the rock-strewn bottom. The lake protected mysterious places too, including four hundred islands, one of which served as a two-thousand-square-foot prison for a now deranged spider monkey, and of course, the largest island, Ometepe.

Ometepe is actually two islands, narrowly joined together, looking like breasts heaving out of the lake. The northern breast is Concepción, an active volcano rising higher than 5,200 feet that still rages to life regularly. The southern breast is Madera, a dormant volcano reaching an elevation of 4,573 feet, the caldera covered with rainforest and an ice-cold lake. While Ometepe is located only five miles from the mainland, in 2001 it still felt rather isolated, covered with its dry forest and rocky beaches, hiding more than 1,400 secret stones carved by Indigenous people, some of whom were Nahuatl-speakers who settled in from the north around 1000 AD.

In many ways, the secret stones define Ometepe well. The petroglyphs are both abstract and representational, zoomorphic and anthropomorphic, and they are carved into the dark, fine-grained basalt that was born of volcanism. Some of the petroglyphs showcase beautiful geometric patterns of unknown meaning that speak to the mysterious quality of the island. Other petroglyphs look like two swirls, carved into rocks now covered in moss, and seem to represent the bivolcanic island itself. Some petroglyphs clearly represent the animals of the island, like monkeys, turtles, and frogs. Others are squared-off representations of people.

While the petroglyphs represent Ometepe well, as an American college student—privileged, with limited perspective and the voyeuristic gaze of a visitor—I thought then that the island was best defined by one man: Rodolfo, a man of many parts. He lived in a dim home with a cement floor and flowery sheets that separated the private sleeping area from the common room. For fifty cordobas, he would give you a jug of rum mixed with ice and some overly sweet strawberry soda. That was Rodolfo the Bartender.

After a few drinks, he would start singing songs about Nicaragua and a young woman who had become lost on the volcano. "Emily,

Emily," he would begin, the name awkward on his tongue, "nunca a re-volver." Then he would continue to strum his guitar with worn fingers, or he might bring his calloused lips to the tines of a plastic sheathed comb. This was Rodolfo the Entertainer.

Rodolfo the Guide was my favorite persona. He patiently led us up the muddy trail, guided us down into the crater of Ometepe's extinct volcano; he helped set up the tent and taught us the Spanish words for the parts of a tree. As evening came, and the mist engulfed us, we began a sorry game of poker. Rodolfo told us his bawdy jokes; we learned every foul Nicaraguan word for the male anatomy and dined on cold rice and beans.

At twenty-one, I had arrived on Ometepe to study herpetology formally, guided by an energetic neuroscientist from a university in New York and his Ph.D.-to-be, tough-as-nails girlfriend. Our class included thirteen students, eight men and five women. One student would go on to become a veterinarian, one a waitress, one a coffee grower in the Caribbean, one a bona fide herpetologist, and another an ecologist. We were bound by a desire to see the world, pursue higher education, and, at a minimum, assuage some curiosity about reptiles and amphibians, aka herps.

That night, two weeks into the program, I sat alone on the dock with my rum, preferring to contemplate the abyss above and below rather than socialize. The other women had returned to their cabin, a wood-planked building covered with green-painted sheets of corrugated metal. Colorful hammocks swung from the eaves and outer posts of each cabin, and wood bunk beds were stacked within. Clothes hung from a line, catching the breeze off the lake. Four or five other cabins, a bathroom and shower building, and a larger main hall—for dining—made up the little academic village set in the dry forest.

This evening, all the women were in their cabin, and the men were split between their cabin and the main hall, where some students were learning how to implant radio transmitters into the body cavities of Northern Boas (now *Boa imperator*). I was ready to leave my lakeside melancholy and go back to the main hall to do the same.

I stood up, the mostly full bottle of amber rum still in my hand. The dockside was dark, but a dim yellow light shone closer to the dirt road that separated the dock and the little academic village. I walked toward the patch of light, habitually looking down and ahead a short distance.

A couple feet in front of me lay a medium-bodied snake stretching two and a half feet long, head slightly raised. The snake sensed my presence. The snake also had beautiful big eyes, dark, thin, vertically oriented pupils, but I knew it wasn't a viper. That didn't mean, however, that its bite wouldn't be laced with toxins.

I stepped forward and leaned down, grabbing the snake right behind the head. I picked it up one-handed and tried to calm it down by wrapping its body around my left forearm, the hand of which was still holding the bottle of rum.

I looked at the beautiful snake more closely, and a little thrill ran down my spine: it was a Cat-eyed Snake (*Leptodeira annulata*). Cat-eyed Snakes tend to live in forests and forest edges, and they're often associated with wet habitat. The habitat I found it in was ideal, an open area along the shore of Lake Nicaragua, surrounded by forest. These snakes are nocturnal, and they hunt for food both in trees and on the ground. I had probably interrupted it in its nightly search for frogs, salamanders, and lizards. This species is rear-fanged, meaning that is has grooved teeth at the back of its upper jaw that help subdue prey and release some relatively mild toxins, at least as far as people are concerned.

This Cat-eyed Snake was gorgeous, long and slender, golden with dark-brown blotches. Its head slightly too big for its body. I was excited to bring it to the main building, a new species for us that many of the students would be eager to see.

When I arrived at the big dining hall, I set my rum down discretely on a table near the entrance. A few guys and the professor stood over a boa at the back of the room. I strolled toward the group, assuming a calm stride despite the adrenaline coursing through my body. I couldn't wait to share my prize, the beautiful Cat-eyed Snake now calm in my hands.

"Hey guys, I caught this near the dock." Two of the guys, who weren't occupied with the surgery, came over. One of them, Javier, was so excited over my find that he grabbed the snake from me so he could position it better and take photos. I felt a pang of loss over the takeover of the lovely snake, but I shrugged it off. I turned to the boa, already on the operating table, and began to pull ticks off his large body with a slender pair of metal tweezers, something we had already been trained to do.

Snakes have an interesting relationship with ticks. They are parasitized by them, like this boa was, and they prey upon them.

It seems that most snakes are vulnerable to ticks, but there is some variation. For example, in one study in northwestern Australia, Spotted and Black-headed Pythons and other species were found with heavy tick loads, but other species with particularly small and snugly overlapping scales, like the Slaty-Grey Snake (*Stegonotus cucullatus*) and Brown-headed Snake (*Furina tristis*), were never found with any ticks. In Australia, other factors seemed to influence tick load too, like habitat and season.

Researchers were particularly keen to investigate the relationship between tick loads and snakes because parasitism can decrease survival and body condition in some species. In the Australian study, no such association was found, but it doesn't mean that ticks don't affect snakes. Some snakes have been documented to experience tick paralysis, which is a type of motor impairment caused by neurotoxins secreted by ticks that have been feeding on a host for a long time. One such case was documented in 2004 in the Florida Keys. In April of that year, a Southern Black Racer (*Coluber constrictor priapus*) was caught at the Key Deer Refuge on Cudjoe Key. The snake itself was lethargic and didn't try to defend itself, and a blood-full tick was found on it. The tick was removed, and half a day later, the Southern Black Racer recovered. This pattern of paralysis and quick recovery when a tick is removed is a classic presentation of tick paralysis seen in other species too.

Even sea snakes are vulnerable to ticks. Researchers have identified an aptly named Sea Snake Tick, a specialist tick that is semi-marine and feeds from the Yellow-lipped Sea Kraits (*Laticauda colubrina*) along the coast of Taiwan. Sea Kraits (*Laticauda* spp.), however, have a much wider geographic range, occurring along the coasts and reefs of the Asian Pacific. The Sea Snake Tick does too. In fact, the Sea Snake Tick isn't the only species of tick that lives on marine hosts, there are also ticks that specialize on Marine Iguanas, Little Penguins, and the Spotted Shag, a cormorant-like bird endemic to New Zealand.

Snakes aren't only the prey of ticks; they are also, in a way, the predators of ticks. University of Maryland researcher Edward Kabay calculated that a Timber Rattlesnake (*Crotalus horridus*) in the forests of the eastern United States could eat 2,500 to 4,500 ticks each year. The snakes aren't looking for a tasty tick meal but instead ingest the ticks when they feed on the mice and other small mammals that make up their diet. Kabay also cited literature showing that increased numbers of mammal predators—like snakes—result in fewer incidences of Lyme disease in people. However, one of our possible saviors from tick-borne disease, the Timber Rattlesnake, is endangered in six states and threatened in five more, meaning that we aren't benefiting as much as we could if their lives and habitat were better protected.

Back in Nicaragua, I continued to pull ticks off the boa for thirty more minutes. Then I watched the deft suturing of the boa's incision site, now packed full with a transmitter, by the veterinarian-to-be. I grabbed the bottle of rum and went to my cabin to sleep.

The next morning, I found that the Cat-eyed Snake had become an overnight sensation. Students were asking where I found it. Others were impressed that I had caught it without being bitten. Javier came up with a look of admiration in his eyes, saying the snake was very aggressive. I realized then that I had been accepted into the boys' club.

Before that moment, I hadn't realized that the boys' club of herpetology existed. I wasn't aware that I had been lacking membership, but it made a difference. Now I was included within the cadre of male conversation in a way I hadn't been before, and I was among the select group of people chased after whenever a new snake was discovered.

The boys' club wasn't meant to be exclusive to men, but admittance seemed to be based on feats of daring and danger, qualities historically considered typical of men. Earlier that week, I had chased after a Speckled Racer (*Drymobius margaritferus*), a gorgeous snake whose blue-to-yellowish scales each appeared outlined in black. That didn't gain me admittance. It was harmless, and I was slow. The Cat-eyed Snake was different. It was aggressive. It was rear-fanged. I had arrived.

The boys' club of herpetologists was strong in the research world at this time, particularly among those studying snakes. In the early 2000s, scientific papers on snakes authored by women hovered between 10 and 15 percent. Today, that figure is closer to 35 percent. A lot has changed in twenty years, and now many of my male peers work closely with women and other underrepresented groups to create a more inclusive field.

Still, among some herpetologists, there is a sense that you are not a "real herper" unless you are diving after venomous snakes or making otherwise docile ophidians perform and strike. This mentality is particularly evident in cable television programs. I call it the "Steve Irwin Effect," and in Nicaragua, I was about to see it firsthand.

The work we were doing on the Northern Boas was slowly advancing our knowledge of snake natural history. In 2001, not many studies had been done that could accurately track the movements of snakes. At the time, automatic data loggers that periodically transmitted an animal's GPS coordinates weren't readily available to scientists. Instead, herpetologists were just beginning to answer the question *Where do snakes go?* with radio transmitters. These transmitters were rather large, about an inch long and a half inch wide, and they needed to be implanted into the snake's body cavity with a surgical operation, like the one performed on the night I had caught the Cat-eyed Snake. The transmitters were large enough that only big, bulky snakes like the

Eastern Indigo Snake, the United States' largest snake species, and other medium-bodied snakes, like watersnakes and hognoses, were candidates for tracking studies.

In fact, in the late 1990s, it was still big news when forty-one Eastern Indigo Snakes had been implanted with radio transmitters at the Canaveral National Seashore, tracked over 1,276 square miles, and still not a single nest had been found. Eastern Indigo Snakes, black and so smooth that they shine with iridescence, used to occur from Mississippi to Georgia, but by the late 1990s, they were only found in southern Georgia and Florida. These beautiful snakes were being poached for the pet trade while their habitat was disappearing. By 2001, the researchers Rebecca Smith and Mike Legare, who had been tracking Indigo Snakes, had now implanted seventy-four transmitters, and they were beginning to discover some worrisome trends. It turned out that a lot of these snakes were found in residential areas, where they were vulnerable to high rates of mortality from being hit by cars. Three years into the study, less than half of the Indigo Snakes that they tracked for more than nine months were still alive.

In the late 1990s and early 2000s, transmitters came in a variety of sizes, with the larger models, weighing around 8.6 grams, tending to have the longest battery life, lasting eighteen months; the smallest transmitters, weighing 1.9 grams, only lasted three months. Protocols were developed to determine what size transmitter a snake got, with maximum transmitter to body mass ratios being about 0.075:1, meaning that a heavy female Northern Watersnake weighing around 400 grams could get a big transmitter, whereas a smaller-bodied male, weighing 100 grams, would need a 7.5-gram or smaller transmitter.

Tracking snakes with radio transmitters during this period taught us a lot about snakes. Once the snake had a transmitter, releasing a particular radio frequency, it could be tracked for months. Some studies taught us about the basic natural history of snake species. Researchers in Ontario learned about the thermoregulatory habitat of Northern Watersnakes, figuring out that snakes obtained their ideal temperature in late morning and that it decreased into the night. Transmitters also allowed for mark-recapture studies through which researchers learned when snakes reached sexual maturity, how many survived as juveniles and adults, and how long a generation was for certain species. This natural history data wasn't just gee-whiz stuff. It helped researchers answer big questions, like what makes a snake species vulnerable to extinction and what happens when snakes are moved away from their territories. Quick answers: slow life histories might make snake species more vulnerable to extinction, and translocated snakes have a way lower survival rate than snakes that stay in their original habitat.

In the study in Nicaragua, once the snakes recovered from surgery, they were released back where they were originally captured, presumably back to their natural home range. A home range is typically defined as the area a snake uses during its daily activities, but this can include both winter dens and active season grounds. Researchers were interested in delineating and calculating the size of snakes' home and activity ranges. They were also keyed up about dispersal—the movement of young as they establish their own home range as juveniles. Sometimes tracking studies were used to uncover even more basic information about snakes, such as the type of habitat they preferred and how much time they spent on the move.

The study I was tangentially involved in allowed researchers to better understand the spatial ecology of Ometepe's Northern Boas by using radiotelemetry to periodically record the location of a snake outfitted with a transmitter. The equipment was bulky, with researchers carrying around large receivers with long, extendable antennas, but it was effective in getting valuable data. This was hard work; snakes need to be tracked by someone in the field, usually found once or twice a day over the course of a month.

At the time, the research that the New York professor was doing was interesting enough to attract the attention of the producers of a quality cable television show devoted to reptiles. You know the type, a show with a guy decked out in safari clothes with plenty of charisma and boldness constantly being filmed wrangling lizards, wrestling caiman, and inching way too close to venomous snakes, alternately yelling as he gives chase or speaking in a stage whisper while sneaking up on some unlucky herp.

This particular TV show herpetologist visited Ometepe with a small crew, including his statuesque producer and a burly, bearded camera man. They were interested in the boas. Our fearless professor, the equally fearless teaching assistant, and a couple of students, including me and the veterinarian-to-be, were filmed over the course of a morning, riding bicycles on Ometepe's dirt roads and lugging around those unwieldy radio-tracking devices in search of Northern Boas.

None of those radio-transmitter-equipped boas were actually found that day, hidden among the boulders and prickly acacias. A local man eventually bagged a boa caught eating chickens out of a coop and brought it to our TV star back at the academic village in the jungle.

At one point, all of the herpetology students gathered in a circle outside the dining hall. The TV-show herpetologist pulled the giant Northern Boa out of the bag and began to describe its natural history. Now at the time, the Northern Boa was thought to be a subspecies of the more commonly known *Boa constrictor*. Their patterns are nearly identical. Today we know that Northern Boas in some populations are often a bit smaller than *Boa constrictors;* they also are known for being a bit quicker to bite.

The TV-show herpetologist shared some general boa facts: they're nocturnal ambush predators, and they typically constrict their prey of rodents, birds, lizards, and frogs. He also brought up human-wildlife conflict. Boas will eat chickens, and chickens are important food sources on Ometepe. In fact, this is one way that snakes were captured for the study. Even Rodolfo would get snakes that—like the one we were circled around—had been caught eating chickens. Rather than killing it as might typically happen, these snakes were brought to the biological research station, implanted with transmitters, and released.

Through all this explanation, the impressively muscled boa constrictor had been calm, but it was time to add that sense of spectacle: The TV-show herpetologist started to manhandle the impressive boa, stretching her out along the ground, pulling her back by the tail. The boa was surrounded, encircled by predators. She lunged, striking at movements made by the students in the circle. One strike made contact with the bottom of someone's hard-soled boot. The thud was sickening. The manhandling continued.

At this point, I left the circle, disgusted by the humiliation of such a proud, impressive boa and disturbed by human beings' capacities to tease and poke and prod creatures with no voice. This was the first time that the shadow-side of herpetology made itself known to me.

Since then, I've seen many herpers showing off their supposed fearlessness and bravery by catching snakes that didn't need to be caught and harassing snakes that didn't need to be harassed. While I've caught hundreds of snakes myself, mostly for research that informs snake conservation and protection, I generally operate by the golden rule: do unto others, including snakes, as you would have them do unto you.

I don't like to be bothered, harassed to the point of striking out, grabbed, or manhandled for other's pleasure. I particularly don't like to be seized by the big hands of potential predators, at risk of having one of my delicate ribs crushed and pushed into my lung or having my spine broken by the heavy pressure of a foot trying to stop me from fleeing or being paralyzed by a careless drop after I've bitten someone in fear.

There's a cautionary principle at work with this empathetic stance, a belief that snakes *feel*. We don't really know what snakes think or feel. That being said, some researchers believe that reptiles do demonstrate basic emotions, including fear, aggression, and perhaps even pleasure. In lizards and turtles in particular, researchers have started to test for basic emotions, but snakes generally aren't included in these studies.

In fact, animal welfare organizations in England worked together to conduct an initial literature review of scientific evidence for reptile sentience. They conclude that it is well established that the behavior of reptiles can indicate their physical health. For example, Ball Pythons often delay feeding when they are stressed or injured. Beyond that, the researchers point strongly to the need for more studies—animal sentience research is an emergent field, especially for reptile taxa.

Since we can't effectively communicate with snakes, or perhaps any other animal, we should play it safe, assuming they are at least sentient, assuming that fear and pain are just as scary and hurtful for snakes as they are for us. This isn't easy though. People just don't see reptiles as being similar to us, and, in general, we don't bond with our reptile pets as strongly as with mammalian pets.

My experience on Ometepe taught me much about herpetology, not as a scientific discipline but as a social construct. I grew into awareness of the strong, aggressive undercurrents of the social structure, of reckless rites of passage, and of childlike disregard for snakes and the way they experience the world.

DISSERTATION

ACROSS THE GLOBE, researchers have docu-
mented declines in snake populations, some of
them dramatic. But the true scope of these declines hasn't been fully explored because we lack
data. For so long, people thought of snakes as
pests to be destroyed. No one worried about them disappearing, and
they certainly weren't concerned about recording population sizes,
birth and death rates, and locations. By the time I began to work at The
Grove, the snakes of Robert Kennicott's day had pretty much disappeared. I was lucky to find a garter snake. I heard tales of a population
of Smooth Green Snakes (*Opheodrys vernalis*) across the road, but I
never found them.

The pain of not being able to experience Robert's faunal bounty in
the prairie, combined with the bulldozing of my own personal snake
haven, the old field I explored with my father year after year, ignited
a desire to fill in the gaps myself. Were there more snakes out there

than I could see? What was really left of those once abundant snake populations? And what factors threatened their future? I applied to a doctoral program in ecology. I studied snakes. In the prairie. In northern Illinois. It was a passion project.

My research hinged on two passions: snakes and the long-lost tallgrass prairie. Like the South American Pampas, African veldts, and Eurasian steppes, the North American prairie is an example of one of the world's most threatened ecoregions: temperate grasslands. By 2000, 41 percent of the world's temperate grasslands had been converted to agriculture; grassland soil is typically nutrient rich and underpins many of the breadbaskets of the world.

The former Canadian Prairies that stretched across Alberta, Manitoba, and Saskatchewan now provide the backbone of Canada's industrial agricultural system, a powerhouse of durum wheat, barley, and oats that exports over $10 billion worth of food to the rest of the world each year. The Highveld of Southern Africa was a region of extensive grassland, maintained by the grazing of wild animals. Today, most of the Highveld has been converted to grain farms or cattlelands, although some South African parks are devoted to its preservation. In Uruguay and Argentina, the Pampas were a vast plain stretching over 460,000 square miles, maintained by frequent fires and grazing herbivores like Coypu, Pampas Deer, Guanacos, and Plains Viscachas—social rodents with colonies a hundred strong. Today the northern Pampas, a grain belt, looks like the middle United States, with crops laid out in squares, bisected by a rectilinear grid of access roads and field boundaries with open stretches grazed and trampled by cattle.

The history of the American prairie is no different. Once prairies covered more than 890 million acres of the central portion of the United States. Today this area is considered to be a critical crisis ecoregion because so much land has been converted to agriculture or grazing and so little land has been preserved.

In North America, the story of the prairies is shaped by major geologic events. In much of the region, glaciers scraped the landscape ten thousand years ago and leveled the land, leaving behind fertile till—deposits of silt, clay, pebbles, and boulders. The legacy of glaciations, along with low precipitation, favored the formation of prairies. The uplift of the Rocky Mountains, finishing up more than fifty-five million years ago, caused an immense rain shadow, where saturated air moving eastward from the Pacific Ocean hits the mountains, rises up, cools down, and falls as rain or snow in the higher elevations, essentially drying out the air before it blows over the Great Plains and gradually begins to regain moisture. This phenomenon creates distinct bands of precipitation and prairie types across the central portion of North America.

The westernmost extent of this range, from Alberta to Montana and south through Wyoming and Colorado to western Texas, was covered by more than 158 million acres of short-grass prairie. The short-grass prairie is more arid than the grasslands of the East and is characterized by delicate Blue Grama Grass, whose comb-like spikes of flowers dominate short-grass remnants each August, along with Buffalo Grass, a wild, sod-forming grass that some folks in California now use for their lawns. This is the land of Burrowing Owls and Short-horned Lizards and rattlesnakes. Today, 85 percent of this region is used for cattle grazing and dryland agriculture.

Moving eastward, one finds a wide band of more than 66.7 million acres that was once the mixed-grass prairie, stretching from Manitoba to North and South Dakota, south through Nebraska down to north-central Texas. The mixed-grass prairie combines characteristics of the short-grass prairie to the west and the tallgrass prairie to the east, largely determined by rainfall. Three-quarters of the region has been heavily altered by agriculture, and virtually no major areas of intact mixed-grass prairie remain today, with a few exceptions in Canada.

The tallgrass prairies, especially the eastern extent known as the "prairie peninsula" that stretched across Iowa, Illinois, and Indiana, has been particularly hard hit. The ecosystem is critically endangered with less than 0.1 percent remaining today. In my home state,

Illinois, prairie once covered twenty-two million acres, and today less than 0.01 percent remains, mostly showing up as little postage stamp remnants.

Most prairies require disturbance to stay treeless, whether that be drought or fire or periodic grazing by native herbivores; Illinois was no different. For millennia, Indigenous people, including the Illini, and after French colonization, the Potawatomi, Kickapoo, Mascouten, Sauk, and Fox, maintained the tallgrass prairie using fire as a tool to support wild game.

That was about to change. In 1818, Illinois became the twenty-first state to enter the Union. At the time, the fifty thousand European American colonizers living there left the prairie largely untouched simply because they didn't have the technology to cut through the dense prairie sod. Then in 1837, John Deere invented a steel-blade plow that would, by 1900, convert nearly the whole Illinois prairie to agricultural land. While the prairie was lost, some of that loss was mitigated by the creation of secondary grasslands, not prairie, but hay-fields or pastures where some prairie fauna could find refuge.

That changed too. After World War II, intensive and mechanized agriculture meant that even those secondary grasslands were lost to row crops. From 1906 to 1987, Illinois lost more than 75 percent of its remaining pastureland. The loss of haylands alone was associated with a concurrent average population decline of 26.7 percent for Illinois's already struggling prairie birds.

The prairie was lost to agriculture, but that wasn't the only threat. Urban areas were expanding, eating up tiny remnants and further separating remaining prairie habitats, as cities' footprints increased by 159 percent between 1960 and 1990. My father experienced this firsthand. Growing up on the north side of Chicago in the 1950s, my dad lived in a 1,600-square-foot, one-bathroom white-clapboard house built in 1906, stuffed to near bursting with four siblings, his parents, grandmother, dogs, and rabbits. With the house so crowded, it was better to be outside, and that's where my dad spent most of his childhood, roaming around what is now the Mayfair Park area. Today, the streets are lined with old bungalows and apartment buildings, but then, some of the lots were still empty and thick with grass, areas that my dad called prairie.

Those prairie remnants, in the middle of sprawling Chicagoland, were the stuff of kids' dreams. My dad would go out to the prairie and dig a giant hole and make a fort. Another little prairie, called Burwell's prairie, "just the size of a small church," my dad said, was a haven for snakes. He'd sneak behind people's houses and flip over wood or tin. Dad said he must have caught every garter snake that ever existed there. The Mayfair Lumber Yard was also a good spot for catching Smooth Green Snakes in the 1950s and 1960s, but go there today and you'll find nothing. Not even the garter snakes, with their tiny home ranges and adaptable diet. The urban landscape just isn't made for snakes.

The more I visualized the loss of the prairie, and the more I imagined snakes trying to navigate the landscape that the once endless prairie had become, the more determined I became to see home through a snake's eyes. I decided to focus my efforts in the Grand Prairie, an area covering most of north-central Illinois. The Grand Prairie historically had twenty-one species of snakes, but the counties I had chosen brought that number down to fifteen, one of which—the Eastern Massasauga—was already listed as a candidate species under the Federal Endangered Species act, and two of which were already listed as threatened in Illinois, the Kirtland's Snake and Plains Hognose Snake.

These two species both were fascinating. In fact, the Kirtland's Snake (*Clonophis kirtlandii*) was first described in the Western scientific taxonomy by my childhood idol, Robert Kennicott, from a specimen in northern Illinois. The snake is small and slender, and Kennicott described it as appearing trigonal in cross section. The snake tends to be brown-backed with dark blotches and a brick-red belly, although one unusual population has white bellies.

The Kirtland's Snake is nocturnal, roaming the night for earthworms, slugs, leeches, and crayfish. It was also known to overwinter in the burrows of its prey, crayfish and perhaps its predators, mammals. This snake is pure prairie peninsula, with its range mostly found in Illinois, Indiana, and Ohio, but it has a penchant for the wet—wet meadows, prairie fens, grassy creeks, pond margins, open swamp forests, and prairie wetlands.

Even as far back as 1892, people had begun noting population declines of the Kirtland's Snake, with one author saying that "tilling, ditching and cultivation of the soil have destroyed [the Kirtland's Snake's] haunts and nearly exterminated it." Being the only member of its genus, with a precipitously declining population, calls for conservation have been loud in some circles. But despite the snake being threatened in Illinois and Ohio, and endangered in Indiana, Kentucky, Michigan, and Pennsylvania, no conservation program exists that specifically targets this relict of the Grand Prairie.

The story of the Plains Hognose Snake is a bit different, and perhaps slightly less disheartening. The first thing to know is that two subspecies of the western hognose occur in Illinois. The first subspecies is the Plains Hognose Snake (*Heterodon nasicus nasicus*), first described in the Western scientific taxonomy in 1852 from a specimen collected by General S. Churchill along the Rio Grande in Texas. The Plains Hognose Snake has a more westerly distribution, found from the Texas panhandle through Kansas north to Manitoba, with a few isolated colonies in Minnesota, western Iowa, and western Illinois.

With a little upturned scale at the tip of their nose (hence the name hognose), perhaps meant to shovel the ground in search of its favorite prey, toads, the tan-and-brown-splotched Plains Hognose Snake is downright adorable. Their burrowing habit means that Plains Hognose Snakes typically prefer well-drained soil that is loose and full of silt or sand but also near water. The sandy sloughs in the western Grand Prairie of Illinois make ideal habitat.

Hognoses are "rear-fanged," which just means that the snakes have enlarged teeth at the back of their mouth that are associated with venom glands. In the case of the hognose, this venom is mild, and its purpose seems to be to subdue toads and frogs, or maybe break down toxins that toads themselves produce. Other than to eat, hognoses almost never bite, although when they do, reactions can vary from temporary pain to extensive blistering. In fact, hognoses have an interesting threat response. When they sense danger, they flatten their heads like puffadders and hiss. If that doesn't work, hognoses then flip over and play dead, with their tongues lolling out.

The other western hognose subspecies in the Grand Prairie is called the Dusky Hognose Snake, and it wasn't formally distinguished until 1952. Together, the Plains and Dusky Hognoses have a pretty large distribution in the middle of North America, but still the tallgrass prairie made up only 18 percent of their historic ranges. In Illinois, hognoses favor sand prairies, but even those have been vulnerable to degradation and destruction, converted to agricultural fields.

With prairie snakes at risk of regional extirpation, I was keen to dive into my research, but I needed to figure out how to capture the snakes first. In my first summer on the prairies of Illinois, I piloted a number of methods. I tried road surveys, traveling up and down the roads at the edges of prairies, looking for snakes—dead or alive. I also tried time- and area-constrained searches, meaning I'd head out into the prairie and walk and walk, trying to find a snake in that tall, thick grass. Not surprisingly, these methods weren't too effective. I found one snake in fourteen days of surveys.

Instead, I had to rely on drift fence arrays associated with funnel traps and coverboards. The drift arrays were made of 1.2-meter-tall silt-fence material, that black fencing, typically at construction sites, meant to stem the flow of water and soil running off the disturbed land. For me, the drift fence arrays—three-pronged, with each prong stretching out five meters—helped guide snakes to my traps and coverboards. For this to work, the bottom of each fence had to be buried. This means that I spent hours—with the help of friends and even my parents—digging nearly a thousand feet of trench into the thick, black prairie sod, pounding in stakes, attaching the silt fence material, stretching that black material across the stakes, stapling them tight, and burying the bottom of the fence in the trenches. This way, a snake that encountered the fence couldn't just go under it but had to slide along it until it encountered a trap or a coverboard and hopefully made itself comfy.

The traps, too, were works of art, handmade by me and my father in the backyard of my childhood home. We made the traps out of three-quarter-inch plywood, cut to make boxes that had a rectangular footprint a bit larger than two feet by nine inches. The boxes were seven inches high and were special in a couple ways. First, the ends of each box were fitted with mesh funnels that pointed inward, ending in two round holes about two inches in diameter. The idea was the snakes could slide along the fence, run into the box that sat along the base of the fence, and be funneled into the trap. The other modification was that these boxes had doors on top, big enough to open, stick both your hands in, and retrieve the snake within. Being concerned about the snakes on the hot prairie, I also put a water dish in each trap, which were a pain to keep filled during fieldwork.

So much of the time, scientists are portrayed as eggheads incapable of real physical exertion or doing anything practical. But the field ecologists I know never quite fit that description. Sure, they are smart and can run code and complex statistical analyses on the computer. Sure, they can pound out journal articles on the keyboard. But they aren't removed from the real world. Field ecologists have to negotiate with tons of stakeholders just to access areas to do their research. They have to spend hours in the field chasing after elusive creatures, enduring the hot and cold, the sun and rain. Half the time, they're digging trenches or digging pits to install traps or equipment. I've had friends who walked up mountains in Colombia catching birds or waded in urban creeks filled with trash and glass. Fieldwork requires grit, it requires dedication, and to be successful, it requires myriad skills—from sawing planks and digging trenches, to negotiating with landowners and calming the power plant manager who didn't get the memo that you're looking for snakes on their property and not intending to blow things up.

For me, every day for stretches lasting two weeks straight, I'd drive a round trip of 240 or 300 miles, stopping at my different field sites, grabbing my field bag full of water, gloves, snake hook, scale, measuring tape, lube, and probes, at minimum, and schlepping it anywhere from fifty yards to three-quarters of a mile to each site. The water was for me and to refill the water containers in each trap. The snake hook was useful when I flipped the tin cover set out at each site and looked for snakes beneath. It allowed me to lift the tin with one hand and gently restrain the snake with the hook until I could bend down and grab it with my hands. The scale and measuring tape allowed me to weigh and take the length of each snake. I also marked them by snipping a little pattern in a few scales near their tail vent. It was like cutting fingernails, and the marks were only expected to last the duration of the study, a few years. With those marks, though, I could tell if I was finding new individuals or the same individuals every time I checked the fences. The lube and probes allowed me to tell if I had a male or female in hand.

To get a robust sample, I set up twenty-two research sites across the prairie preserves of central Illinois. The easternmost preserve was Midewin National Tallgrass Prairie, the first national tallgrass prairie preserve in the United States, stretching across nearly forty thousand acres of agricultural fields, rangelands, forest, and wet prairies. Of the forty thousand acres, fewer than five hundred acres were considered to be high-quality prairie, and this was still disturbed.

For many years before I started my research, Midewin National Tallgrass Prairie called to me. I imagined acres of floristic wonder: Compass Plants with stalks of yellow flowers reaching toward the sky, Purple Coneflowers dancing like Degas's ballerinas in soft, pastel purple tutus, Big Blue Stem coloring acres purple and green. My imagination was an Impressionist painter's dream of the tallgrass prairie. Reality was much different.

My first research site at Midewin was located near huge, humped bunkers. These bunkers were built in the early 1940s on what was once U.S. Army Ammunition Plant property. During World War II, the munitions factory produced artillery shells, antitank mines, TNT demolition blocks, bombs, and a billion pounds of high explosives. The dome-shaped bunkers allowed explosions to move up and not out and comprised a complex of more than a thousand buildings and 164 miles of railway. The munitions factory was like a city, with a U.S. Army Health Clinic, a director of plant operations, deputies of procurement and production and material management and maintenance, and ten thousand employees. This city would get a bit sleepy after the end of World War II, but it woke back up during the Korean and Vietnam Wars, finally closing in 1976. Twenty years later, the land was transferred to the U.S. Forest Service.

The name Midewin itself is derived from a Potawatomi word connoting healing, and that is exactly what the land was doing. It was recovering its garb of native flora and fauna while reconstituting a region of the country that had lost too much of its natural heritage. Midewin wasn't pristine, and it wasn't the untouched prairie of the imagination, interwoven as it was with cow pasture and former agricultural lands, but it was still incredible. The huge prairie preserve was home to at least 570 plant, 172 bird, 53 fish, 27 mammal, and 15 reptile species. In some areas, restoration has been intensive, with rows of native prairie plants being cultivated to provide seed for rigorous replanting efforts. The more than eighteen-thousand-acre reserve reintroduced state endangered flora, like the Royal Catchfly, and years after my research was complete, it established a conservation herd of American Bison. These efforts have been the result of hard work and fruitful collaboration among the Forest Service, the Nature Conservancy, and other nonprofits including the Wetlands Initiative, Openlands, and the Chicago Wilderness collaborative.

While my research sites didn't have unobstructed views of endless prairie, they were still monuments to resilience. At Midewin, in the summer, the sky typically is a pale blue, never bright and crisp like it is in the western United States, but soft and gentle, spotted with cottony white clouds. If you stand in the middle of a field, grass extends in all directions, but it's not endless. In the distance, you can spy a barbed wire fence, along which, at intervals, you'll find an Osage Orange, some invasive shrub, or a young tree. At some point, the viewshed is interrupted by a farmhouse, which always registers white to the eye, and a longer stretch of forest, where the land slopes into the floodplain of a river or stream.

In June, the fields are a bright, fresh green with a bit of white Daisy Fleabane poking out here and there. The green at Midewin isn't always bluestem and Prairie Dropseed, but extensive fields of Timothy, Orchard Grass, and Hungarian Brome, with their seed green and growing. By the end of July, the grass is dark green and starting to brown in patches. Some strange beauties emerge, like Horsetail Milkweed, with creamy white flowers, architecturally arranged on top of a stalk with dense, slender leaves. In mid-August, the less pristine grasslands are white with the flat, flowering cymes of Queen Anne's Lace; the grass is starting to brown, and the purple flowers of clover peak out in areas where the grass is thin.

In these large, relatively roadless tracks of partial-prairie or former pasturelands, I was able to find snakes that felt like home; snakes that were comfortable and handleable; snakes that had adapted to the bison and the fires set by Indigenous peoples, to the colonizers first from the eastern United States and then immigrants from Eastern Europe, to the grazing cattle and noise of munitions plants. I found Common Garter Snakes, Plains Garter Snakes, Dekay's Brown Snakes, Blue Racers, and Eastern Fox Snakes too.

The fox snakes were rather a treat, being long-bodied, stained walnut-brown with coffee-colored blotches, and having very kind faces that seemed to smile. In 1853, an Eastern Fox Snake specimen collected by Dr. Philo R. Hoy in Racine, Wisconsin, was used to characterize a species new to the compendium of Western knowledge, but not, of course, to the Indigenous peoples of the Midwest. Dr. Philo R. Hoy was a redheaded medical doctor cum ornithologist cum archaeological preservationist, nicknamed "red-headed woodpecker." As Dr. Hoy made his rounds in Racine, he usually carried a pocket lens, butterfly net, and glass bottles for collecting insects. Dr. Hoy's fascination with the natural world resulted in an impressive collection of 318 bird specimens mounted by him and his wife, Mary, and a record of thousands of insects and moths. Dr. Hoy also recognized the important heritage of Racine's Indigenous peoples and worked against the odds to preserve fourteen Indigenous burial mounds. With an expansive mind and vigorous energy, he served as a state commissioner of fisheries and was a member of the Wisconsin Board of Health, the Geological Survey, the Entomological Society of France, and the Chicago Academy of Sciences. He also served on the Professional Board of the Smithsonian Institution, which means that he knew how to get a fox snake that he found on his ramblings in Wisconsin to Washington, DC, for scientific description by Spencer Baird and Charles Frederic Girard.

Like Philo Hoy, Spencer Baird and Charles Frederic Girard were fascinating and energetic mid-nineteenth-century scientists. Spencer Baird established himself as a naturalist in Pennsylvania, having trained informally with John James Audubon and others, before dedicating his life to expanding the Smithsonian. As the first curator for the Smithsonian, Baird donated his personal collection of specimens—two railroad boxcars full—to the museum, and then set to work developing the museum's natural history collection. Baird also mentored young naturalists around the country, including Robert Kennicott. Today, The Grove National Historic Site contains replicas of cabinets designed by Baird to house Kennicott's own natural history collection.

Charles Frederic Girard, who codescribed the fox snake with Baird, was a Frenchman by birth. Girard had studied under the renowned Swiss natural historian Louis Agassiz and later came to work at the

Smithsonian as Baird's principal assistant. For a decade, Girard published species descriptions of herpetofauna and fish at the Smithsonian, while earning a medical degree from Georgetown. Then, in the early 1860s, Girard accepted a position with the Confederacy, poised for war, as a supplier of drugs and medical supplies from France. At this point, Girard abandoned his work at the Smithsonian and dedicated his life to medical practice, serving as a military physician during the Franco-Prussian War and living the rest of his life in France.

Baird and Girard's description of the fox snake forefronts their combined expertise and dedication to scientific objectivity but fails to enliven this delightful species. "Head rather short, vertical broader than long," they say, rather than noting that the shortened head gives the Eastern Fox Snake a rather friendly countenance. "A series of broad transverse quadrate chocolate blotches extending from head to tail, about 60 in number, 44 to anus" is a rather apt description of the dark-brown patterns that stamp the back of this light-brown prairie snake, and I like to think that the use of the word "chocolate" indicates a latent love for the species.

In fact, the fox snakes are rather complicated. The fox snakes of central Illinois were considered to be Western Fox Snakes by most authorities until 2011, under the name *Elaphe vulpina vulpina* and then *Pantherophis vulpina*. This was a group of fox snakes whose range stretched from eastern Indiana across the Mississippi River to eastern South Dakota and Nebraska. Another subspecies of fox snake, *Elaphe vulpina gloydi*, later considered its own species, *Elaphe gloydi*, slithered around the edges of Lake Huron and Lake Erie from northern Ohio up. But things change . . . at least they do in the world of snake taxonomy. In the case of the fox snakes, those in Illinois are now considered Eastern Fox Snakes (*Pantherophis vulpinus*), and the dividing line between the species, according to DNA evidence, is the big geographic barrier of the Mississippi River.

Taxonomy—that is, classifying animals—has always been complicated and rife with polemics. This stems, in part, from the fact that taxonomy is actually important. It is fundamental to any discussion on natural history because scientists need to know what organisms they're talking about. It also has real-world consequences—you have to exist as a species (or subspecies or a distinct population segment) in order to be considered, say, for the endangered species list. This also stems from the idea of taxonomic freedom, which acknowledges and accepts debates on how organisms should be classified, with different folks giving different weight to major geographic boundaries, population-level variation, and variations in DNA (that is, molecular phylogenetics). That being said, taxonomy isn't some slapdash affair.

Recently, the way Western scientific taxonomists have delimited species of snakes—and reptiles and amphibians more generally—has shifted. Historically, species were identified based on morphological differences and geographic factors, but in the 1960s, advances in the field of genetics allowed researchers to formally test for reproductive isolation by looking at gene flow. By the 1990s, the use of DNA exploded with methods that allowed for the "barcoding" of species. With these methods, researchers were noticing more and more differences among species, leading to a huge increase in new herp species, mostly a result of splitting old species into many distinct new ones based on the new DNA techniques. But some researchers have pushed back on this trend. David Hillis, from the University of Texas at Austin, wrote that "inadequate sampling and a lack of attention to contact zones often leads to the over-splitting of species," and he advocates for a more formal combining of the two techniques, those used by taxonomists of old that focus on morphology and geography and those used by taxonomists today that focus on DNA.

To make things even more confusing, sometimes snakes hybridize. Hybridization usually occurs within a genus, for example, a *Pantherophis obsoletus* (Western Ratsnake) and *Pantherophis guttatus* (Corn Snake) might interbreed and result in fertile offspring. Sometimes hybridization can occur between snakes that are quite different.

In fact, intergeneric hybridization—or hybrids between two different genera—have only been reported among snakes three times in the literature. One of these intergeneric hybrids was reported in the 1940s, between a Timber Rattlesnake and a Massasauga, and was based solely on how hybrid offspring looked. Another, reported in 2009, was based on a photo.

Only recently, in 2012, has DNA been used to confirm an intergeneric hybridization, and this was between Bullsnakes (*Pituophis catenifer sayi*) and Western Fox Snakes (now *Pantherophis ramspotti*) at sites in southeast Minnesota and south-central Iowa. The hybrids looked and acted differently too. The hybrid in Iowa couldn't quite hiss as long as the Bullsnake, known for their ability to huff and puff. The number of scales on the hybrids was intermediate between the two species, which typically have a set scale number. Photos of the two hybrids are fascinating. Like a Bullsnake, the two hybrids have black markings on the scales around their mouth, but the rest of the patterning on the head is much reduced, like a Western Fox Snake. The head shape of the two hybrids is intermediate between the elongated, pointed snout of the Bullsnake, and the cute, round head of the Western Fox Snake. The hybrid, really, is quite beautiful.

Whatever we call them, fox snakes, like their congeners the Black Rat Snakes, have been observed in ritualized combat, vying for a mate. With a little less than 66 percent of their current range coinciding with the historic tallgrass prairie, fox snakes are firmly situated in the web of grassland life, consumed by birds of prey and consuming rodents and rabbits. During my research, though, I only found these fox snakes at the sites with the most acreage and the fewest roads. That indicates that fox snakes are sensitive, like so many large-bodied snake species, to habitat fragmentation.

I also visited another site, Sunbury Railroad Prairie Nature Preserve. Sunbury was a floristic dream. Beautiful golden-headed *Silphium* species stretched toward the sky; delicate gentians bobbed their bell-like heads; and splendid, purple *Liatris* stood erect, like guards over the last remnants of prairie in Livingston County, Illinois. The problem was, this site was wedged between agricultural fields and roads. Covering an area of only twelve acres, stretching one mile, and sitting not quite one hundred yards wide, it was a postage stamp preserved because it was once a railroad easement. As beautiful as Sunbury was floristically, the surrounding landscape was devastating to snakes, and I only found one individual snake there—another lovely little fox snake.

While I captured 120 snakes across three summers and twenty-two sites spread across six tallgrass prairie reserves in northern Illinois, the stories of two preserves, Midewin and Sunbury, are enough to understand the current status of and challenges faced by tallgrass prairie snakes today. Looking at historical data, it was clear that since the 1930s, midwestern snake populations had declined.

It also became clear that even though there were stark differences in the floristic diversity of the sites—with my particular sites at Midewin being largely degraded floristically and Sunbury being rich in a variety of forbs—these microhabitat differences hardly registered in terms of snake abundances and diversity when compared to the impact of landscape.

Moreover, clear gradients in snake composition were seen as we moved from the rural west to the more urbanized east. Garter snakes made up over 75 percent of my captures, and they had high abundances in areas with high urban cover—think small plots, lots of roads. This makes sense because even though garter snakes are fairly large, they have very small home ranges. They are dietary generalists and can spend time in a small area to get their needs met. This means that the garter snakes don't have to interact much with the surrounding landscape. They're not moving across roads, dodging cars, like those bigger snakes with larger home ranges—the fox snakes and racers.

Those beautiful Blue Racers had quite an aversion to human landscapes, always more abundant the farther they were from an "anthropogenic edge," whether that be an agricultural field or a road. Fox snakes were more abundant in areas where grasslands were expanding rather than contracting. The landscape is critical to snakes. They have no way to avoid it, no way to evade the hazards the surrounding landscapes present. They can't fly over roads or even run across them quickly. They can't jump out of the way of a tractor or plow. Snakes are in the thick of things, and everything we as humans do, they experience the consequences of fully. That is, if snakes are still around us at all.

The challenges snakes face were made more concrete when I was out on the wet prairie one morning. The rock-gray clouds hovered low overhead, slowly parting to expose some blue-sky freedom. A recent rain shower had left muddy puddles in the gravel road. I was leaving a field site in my blue jeep, driving through the craters in the road and letting the water rise in thick waves along either side of my car.

Something caught my eye: a thick, rubbery C-shape in the puddle ahead. I stopped the jeep in the middle of the road. In the shallow pool lay a stout, blue-tinged Plains Garter Snake. Her fat belly had burst under the weight of a car, split down the middle. Radiating from the burst belly, like the rays of the sun, were twenty slender, three-inch-long babies. All dead.

I remember my stomach heaving. I remember my heart bursting like the blue-sky freedom pushing through the gray clouds. I remember getting back into the jeep. The image of the exploded mother snake surrounded by her young blocked my view of the road. For days, the image was there. For days, I drove those gravel roads slow and steady and with dread.

I brake for snakes.

In the mid-1800s, a dead snake, even a dead snake with twenty dead babies, would not have concerned most prairie-hardened farmers in the grassland provinces of Illinois. If accounts are accurate, the region was "awfully thick" with snakes.

M. F. Lawson, an observer in Warren County, Illinois, in 1841, describes what for many would be a horrifying scene: "Eels are not known, but snakes are, to the extent to supply all deficiencies. It is an excellent precaution, when going to bed in the dark, to take the bedclothes off and shake the snakes out of them before getting in yourself."

In those days, Illinois's thirty-seven snake species were so abundant that, when settlers first broke that thick prairie sod, killing snakes—those "big fat fellows"—became sport. Harry Eenigenburg, born to Dutch immigrants on the Illinois prairie, described farm boys gathering and killing "four or five hundred snakes of all kinds in one drive."

Another settler, William S. Pearse, described the typical method of killing snakes: "A stroke with a switch or whip breaks their joints and disables them and then it is the custom of the country to put the foot on their heads, catch hold of their tails and pull their heads off."

In 1843, Margaret Fuller, a Massachusetts journalist and American Transcendentalist, toured the Illinois prairie. She waxed poetic about the expansive prairie "continually touched with expression by the slow moving clouds."

Fuller also rang a death knell for snakes, explaining that "wherever the hog comes, the rattlesnake disappears; the omnivorous traveler, safe in its stupidity, willingly and easily makes a meal of the most dangerous of reptiles." Settlers had introduced domesticated animals that could make quick work of snakes, including the diminutive Massasauga rattlesnake.

After World War II, the character of the Illinois landscape changed. The endless expanse of six-foot-tall Big Blue Stem and star-flowered Compass Plants had long been decimated. Mechanized agriculture destroyed even the secondary grasslands that remained. In Illinois's northern Grand Prairie counties, acreage in hay declined 51 to 100 percent between 1957 and 1987.

With the loss of the prairie and secondary grasslands, the most sensitive snake species in the tallgrass prairie largely disappeared, including the Bullsnake and Massasauga rattlesnake, which today borders on extinction in the prairie state. Adaptable species, like garter snakes, could still be found on small acreages. In fact, in 1947 two researchers—one from the University of Illinois and the other from Indiana University—found 383 snakes in a 3.2-acre field south of Chicago.

By 2007, counts of snakes in the once "grand" prairie of Illinois were downright dismal. Intensive agriculture, and the invasion of roads, proved too much for most snake species. During three summers of research on the prairie, I found fewer than 44 percent of the snake species that settlers used to shake from their bedclothes. More than two-thirds of the individual snakes I did see were garter snakes, but even their numbers have drastically declined. With intensive collection over many months, using the most effective known capture methods, I only captured a third of the number of snakes that researchers had sixty years earlier.

This is cause for concern. Snakes are a fundamental part of many ecosystems, acting as both predators and prey. And the loss of individual snake populations or entire species could have numerous unforeseen consequences. Considering extinction from a promontory, Thomas Jefferson articulated the risk that species loss poses: "For if one link in nature's chain might be lost, another and another might be lost, till the whole system of things should vanish by piece-meal."

While the Illinois landscape of Jefferson's time is unrecognizable today, the attitude of most of the state's tough farmers has not changed. Another morning, I was driving the graveled roads in the wet prairie. I pulled over to the side to move a garter snake off the road.

A big truck pulled up beside my car. A man with a rock-gray cap eased himself out of the cab. He asked, "What have you got there?"

I showed off the snake like a newborn baby and told him it was a Plains Garter Snake. I explained that I was moving it out of the road. As we stood under that blue-sky freedom, he looked at me, his eyes narrowed. "Huh," he said. "You know, the only good snake is a dead snake." He turned around and got back into his truck without a backward glance.

My stomach heaved. My heart burst again, and I sank to my knees. There, in the weedy remnant grasslands that border dirt roads, I mourned the loss of the wild prairies. I mourned the loss of the wild snakes. And I mourned the loss of the blue-tinged Plains Garter Snake and her twenty slender babies.

VULNERABILITY

I REMEMBER THE FIRST time I visited Lake Waccamaw, a large Carolina bay lake covering 8,938 acres with fourteen miles of shoreline. I was twenty-four, living in the Piedmont of North Carolina, and I drove a few hours to experience a new geography closer to the coast. I was mesmerized by the tea-colored water, dyed with tannins from the adjacent swamplands. I was entranced by my own imaginings of the region a millennium ago. How tall and thick were the looming cypress trees then? How many alligators could one see in a single day? Tens? Hundreds? I was hypnotized by the landscape, by the lapping waves, hardly able to pull myself away from the shore, but I did, and I began to walk.

At first, each step through the shrubby bay forest elicited a dry crackle that I feared would frighten away the Northern Parulas I heard buzzing overhead, but they didn't seem to mind. Even the vivid-green Carolina Anoles, hanging onto the smooth bark of a sweetbay, barely seemed to pause as I walked by.

I walked along the shore toward Waccamaw Creek. The trail wound through forest already desiccated in late May. The warm breeze made the blushing blooms of Rose Spiderwort dance. The grass rustled a few feet ahead of me—and I caught a glimpse of a long black tail. I rushed ahead, trying to step quietly, and I was rewarded by the sentient stare of a Black Racer. Our eyes locked for a few moments, until it slipped away into the swamp forest.

My senses were alive now, my snake vision activated. My eyes focused about eight feet ahead, scanning from side to side, waiting to catch a glimpse of the next snake. Minutes later, I spotted another rubbery, black crescent ahead where some small trees were growing next to the water. It looked like a piece of shredded tire, another Black Racer warming itself in the dappled sunlight.

The hike continued, my search continued, and only four feet ahead, my eyes converged on a coppery, semi-coiled form in the dry grass. I walked slowly, hunching slightly, stopping every couple steps. The pattern of this much-maligned ophidian was bewitching: salmon pink mottled with bronze. I got closer, focusing on the copper eye and black slit of the pupil. This Copperhead was savagely beautiful. I was tempted to take one step closer. I closed my eyes for a second and then, reluctantly, took two steps back and continued on my way.

Now I was walking deliberately, always searching, always hopeful. The hunt was addictive. The path widened, covered with dry brown leaves. A bright sinusoidal shape sharpened into focus. Another Copperhead perhaps? This snake was long, though, and comparatively thin-bodied. I rushed ahead—a Corn Snake! It retracted into an exaggerated S-shape, its upper body held above the ground revealing a perfect checkerboard pattern on its belly. The contrast between the bright base color of the back, splotched with dark orange, and the belly, with every second or third line of scales shifting from black to white to black again was incredible.

By the time I reached the dam that flooded Waccamaw Creek and began the hike back to my little blue jeep, I had seen twelve snakes of four different species—Northern Black Racer, Southern Copperhead, Red-bellied Watersnake, and Corn Snake. I thanked the snake-hunting gods above, greedy for more and wondering what my next adventure would yield.

But adventures like these have grown more infrequent, with fewer snakes gracing my walks each year, and that has everything to do with the interaction between snake biology and human behavior.

Snakes are exquisitely vulnerable. They lack legs and wings and thus stay in full-body contact with their environment. Snakes are intimate with their landscape in ways that other animals simply are not. They cannot fly between their breeding grounds and winter habitats; instead, they must slide through the matrix, the habitat in which they're embedded, bellies to the ground.

Today, that matrix is filled with danger. In cities and suburbs, snakes are killed by cars, killed by people who loathe them, and killed by nonnative predators, like cats, and mesopredators, like raccoons, whose populations have skyrocketed in these human-modified landscapes. The agricultural matrix is no better. Agricultural equipment tears at snakes and their eggs, and pesticides and herbicides harm snakes too.

In relatively rural areas, the matrix is still damning, with even low-traffic roads killing off vulnerable populations. Around the country, researchers have documented the effects of road mortality on snakes. At Carlyle Lake, in Illinois, one of the last strongholds for the Massasauga rattlesnake in the "corn desert," forty-two Massasaugas were recorded dead on the road in just thirty-two months, and most of these deaths occurred in August and September, when peak mating season and peak tourism season collided.

In the sagebrush steppes of southeastern Idaho, the story is little different. There, Gopher Snakes made up nearly 75 percent of the road observations, with peak mortality happening twice a year, in the fall and spring. The researchers supposed that Gopher Snakes were seen more often than rattlesnakes or garter snakes because of their ecology—they are active rather than passive hunters; they move more than rattlesnakes, which sit and wait to ambush their prey. Plus, Gopher Snakes are long-bodied, making them easier targets that motorists were seen purposely swerving to hit. In the Everglades, too, lots of snakes are killed by cars, and the peak mortality has an ecological correlate, in this case, seasonal fluctuation in surface water levels corresponding with snake migrations.

The detrimental effects of road mortality on snakes, however, isn't just limited to the United States. In Europe, juveniles of an isolated population of Bohemian Aesculapian Snakes are disproportionately impacted by road mortality. In Australia, nearly 10 percent of the snakes seen during one road survey study were dead. In a protected rainforest in Brazil, sixty road-killed snakes were found in fourteen months. This may not be a lot compared to the number of snakes in the area, but as the researchers rightly point out, given this is a protected area, measures should be taken to protect the fauna, including snakes. In Taiwan, an innovative citizen science project undertaken between 2006 and 2017 resulted in more than eleven thousand records of road-killed snakes, mostly in areas of good habitat.

Snakes' road mortality has been documented for decades, but the problem doesn't seem to be going away. Researchers predict that more than twenty-five million kilometers of new roads, added to the sixty-five million that already exist, will be constructed by 2050, and much of this expansion will be in Asia, where herpetofauna populations are already threatened.

The effects of roads on snakes are complicated. Yes, roads can kill them, but studies completed around the world make it clear that, depending on the snake species and its ecology, the effects of that mortality can vary widely. In some cases, adult males seeking a mate are disproportionately killed by cars; in other cases, the dead are dispersing juveniles. The overall effects of this mortality on snake populations can vary just as much, perhaps most affecting snake species that are the longest lived, with slow population growth and late reproduction. Roads don't just outright kill snakes; they also can limit the size of home ranges and shift the types of habitats that snakes are occupying, which can slow growth, delay sexual maturity, and lower reproduction.

For snakes, the world can be a scary place. Around the globe, researchers have been documenting declines in snake species. Habitat destruction and fragmentation, mechanized agriculture, illegal harvesting and overharvesting, and invasive species all threaten snakes. Plus, there are possible threats out there that haven't been well researched. How might climate change affect snake populations? How does the increase in mammalian mesopredators—like raccoons and domestic cats—affect the remaining snake populations?

Even seemingly innocuous objects have big consequences. For example, snakes are vulnerable to landscape fabrics, especially rolled products with net-like mesh made of plastic or nylon. These products usually hold something down, like straw or jute or wood, allowing grass seed to germinate or preventing erosion at sensitive ecological sites. Unfortunately, these products can be deadly. Researchers in South Carolina have found at least five snake species dead and tangled in erosion-control fabric, including Black Racers, Rat Snakes, watersnakes, Corn Snakes, and Eastern Hognose Snakes. I've had my own experiences with this.

Each year, I go camping with my family at a small state park in eastern North Carolina, where we set up our tent against a dark, cypress-filled swamp a short walk from Lake Phelps. Lake Phelps is remarkable, stretching 16,600 acres. It is one of about a half million elongated, oval swamplands with puzzling origins that dot the eastern coast of the United States from Delaware to Florida. Most of these are found in the Carolinas, where they're called Carolina bays.

Legend has it that researchers weren't even aware of the bays as an unusual phenomenon until the 1930s, when black-and-white aerial photographs were first taken of the region. The photographs revealed a series of elliptical depressions, some in the middle of farmland, some as marshland, and some as open lakes, oriented in a northwest-southeast direction but rotating northward as you move from Georgia to Virginia. This rotation also matches the change in direction of parabolic or U-shaped sand dunes in the same area that formed during the late Pleistocene, when much of the northern United States was covered by the Wisconsin ice sheet.

There's evidence that the Carolina bays were also formed in the late Pleistocene. Radiocarbon dating indicates that the lakes are at least thirty-eight thousand to fifty thousand years old. Analysis of sediment cores show pollen consistent with plant species from the late Pleistocene and early Holocene.

Despite having a good sense of *when* these lakes were formed, the *how* has remained controversial. In the early years after the mysterious lakes were photographed from the air, some scientists suggested that the depressions were formed by impacts from asteroids and comets. It was only in 2009 that scientists firmly rejected that hypothesis due to the shallow depth of the lakes and the lack of meteorite fragments. More mainstream hypotheses suggest that water or wind created the Carolina bays, perhaps formed by currents when the area was under the sea and then further hollowed out by winds during the Wisconsin glaciation, modifying their shape and giving them a similar orientation as those late Pleistocene dunes.

Regardless of how the Carolina bays were formed, they remain ecological wonderlands, rich in biodiversity. Some bays are lined with unusual carnivorous plants, like sundews and pitcher plants. Other bays are rimmed with Black Gums and Bald Cypresses, sweetbays and pawpaws providing rich habitat for egrets, owls, bears, and herps.

Several years ago, in early May, my family and I pitched our tent within a hundred yards of Lake Phelps, near a hollowed-out Sycamore tree and tucked in among looming Bald Cypresses. At night, we would hike empty trails, watching opossums and raccoons scurrying about the swamp, and basking in the serenades of the Southern Toad. Sometimes we would stop for long moments to watch fishing spiders and share a moment of interspecific stillness or to mourn the death of an American Bullfrog, flattened on the pavement by a car earlier that day.

We hiked our favorite trail the next afternoon, one that passed through deep, old woods that skirted the lake. The birds here were magnificent, and I caught the songs of Great-crested Flycatchers and Wood Thrushes among the more common denizens, the Red-bellied Woodpeckers, Carolina Wrens, and Northern Cardinals. Frogs would sometimes interrupt the repertoires of these birds. A Green Frog would start its banjo twang, or a Southern Cricket Frog would start snapping crisply, the sound of two marbles hitting each other.

The trail was a feast for the eyes too. Zebra Swallowtails, looking like heavily striped white tigers, flitted among the pawpaws, which still carried some of their mysterious maroon blooms, perfectly shaped, with six petals hanging facedown from trees barely leafing out. Diminutive brown-and-black Ground Skinks scurried into the leaf litter as we walked past. Blue tent caterpillars, backs hairy and covered with white, vase-like splotches and orange stripes, wriggled and curled around twigs.

Soon, we approached one of the best spots on the trail, a section bordered by a rivulet of tannic water heading into the lake. The trees between the trail and the stream were young, with slender trunks, making it easy to peer across to the other side to catch sight of a box turtle or snake warming itself in the dappled sun. We rushed up to the popular sunning spot, but this time it was empty. We continued on, senses sharpened, looking up into the trees for birds and down to the ground for herps.

Then we saw our first snake. I saw it from a distance, a Copperhead. Something didn't look right. As we moved closer, we saw that the Copperhead was alive, but stuck in landscape mesh laid down to reduce erosion into Lake Phelps after a small canal weir had been repaired. As we looked across the two-hundred-square-foot area that had been covered by wide mesh and hay, we realized there were more. They were hard to see. We saw a Red-bellied Watersnake tangled at the edge where it had found a sunny patch near the canal bank to warm up. We found a Black Rat Snake struggling and pulling, the mesh only tightening and cutting deeper into its skin.

And then we sprang into action. I had a first aid kit in the car; my husband ran back up the trail to get it and brought it back. I tried calling the park ranger but received no answer, and my son and I found at least five more snakes, two of which were already dead. Then the three of us worked together, starting with the nonvenomous snakes, those that looked the most injured. I would secure their head, and my husband would carefully snip at the mesh and free the snake. He'd hold the snake out, examining it carefully, as veterinarians like himself are wont to do, directing me on where to put the antibiotic ointment when he had seen a spot where the mesh had sliced past the snake's protective scales.

In less than a couple hours, we freed six living snakes of five species—Red-bellied Watersnake, Brown Watersnake, Eastern Rat Snake, Black Racer, and Copperhead. The snakes we saw that day had been the only ones we had seen at Lake Phelps all weekend. When our work releasing and treating snakes was done, we rolled up the mesh and hay, angry at what had happened. The park was supposed to be a rare place of refuge for these animals.

Then we dragged the bundles behind us and marched over a half mile up the trail. Each of us loaded like a sled dog, heaving and huffing in the intensifying heat of the afternoon. We dumped our load at the side of the road and drove to the park ranger office, half-filled with indignation at what we had just seen and half-filled with fear that the ranger would yell at us for tampering with their erosion-control fabric. There was no need; the ranger was receptive and put the snakes first. The fabric had only been put out the afternoon before, and already eight snakes had been caught. The ranger was as appalled as we were.

I cherish these moments, these small moments where I feel I'm making a difference in the real world, but my friend Simon, considering the massive scale of habitat loss and animal death, once asked me, "Isn't it just a drop in the bucket?"

All I could say is, "Yes."

Some people focus on a narrow mission for which they're filled with purpose and fervor. Often, this type of energy goes into for-profit enterprises, such as Steve Jobs's single-minded focus on building a "computer for the rest of us." Research suggests that when we devote our energy to small, focused projects, like stopping the spread of white-nose syndrome or saving Mole Kingsnakes or preventing bird-window collisions, we can start filling a small bucket. Many of us could do that—devote ourselves to one thing and make a tremendous impact in a small domain.

That model is top-down, based on one person's vision. And their vision needs to be passed to others through persuasion, advertising, even coercion. My own mode, however, is more traditionally feminine, a mode of presenting information, creating personal connection to a topic, and letting others choose their perspective, freely and without pressure.

As a teenager, I patterned my understanding of influence and power on the ideals espoused during the American Enlightenment. It was unseemly, then, to actively seek power—Aaron Burr was criticized for campaigning openly, and George Washington was applauded for giving up his power after two terms in office. But maybe power isn't the real problem, just our models of it.

In the United States, our most exalted models of power involve coercion, force, and even death: Washington as warrior, Washington as enslaver. But as Elizabeth Lesser proposes, there are other ways to wield power: we can create partnerships, be interactive, forge connections, value relationships and empathy and communication, be generous with praise, open up to vulnerability, and listen. This stands in contrast to relying on hierarchies, being authoritarian, forcing competition, valuing individualism and fortitude, withholding praise, denying mistakes, and overriding the views of others.

As an ecologist, it's also hard to devote myself to one aspect of the natural world because it is profoundly interconnected. Ecology is perhaps the most interdisciplinary and connection-focused science. It's fueled by questions that bring in other organisms and biotic processes and geology and human land use history and values. Ecologists can't see an issue like bat species declines or herpetofauna declines or bird-window collisions without seeing those connections. That means that it is hard to commit to a single, specific, measurable goal—like ensuring cavers everywhere disinfect their gear before entering new caves or stopping land conversion to save the Kirtland's Snake in midwestern wet prairies or ensuring the top-ten major cities in the United States develop effective regulations to prevent bird-window collisions.

If I could devote myself wholeheartedly to any of those goals, I'd be able to make a measurable impact on one thing. I'd see a bucket filling.

The problem is that my real goal is broader: to help my fellow creatures, human and nonhuman, thrive, to help them live lives of meaning, value, and success, and to help people develop a positive connection to the natural world that might prime them to protect it, respect it, and grow with it.

A business consultant, an essentialist like Greg McKeown, would likely say that this goal is too big and too vague. And I see that. By pursuing this goal, I'm putting drops in the "spiritual connection to nature" bucket and drops into the "intellectual connection to trees" bucket and drops into the "habitat of snakes" bucket. None of those buckets will ever be filled by the fruits of my effort alone.

Am I misguided? Have I wasted my potential to make a difference by having such a broad aim? Are my efforts adding up to nothing? Did I pick the wrong idols? What was Robert Kennicott's goal? He wrote articles for the local paper. He collected snakes. He traveled to Alaska. He isn't widely known. Did his work really matter? What about Donald Culross Peattie? What was his goal? He wrote. He wrote about trees and naturalists and a little town in France. He isn't widely known. Did his work really matter?

I can't help but feel that something is missing from discussions of "goals" and "SMART objectives" (SMART serving as an acronym for Specific, Measureable, Achievable, Relevant, and Time-bound). I can't help but feel that crucial connections are being snipped. Shouldn't SMART objectives have other criteria? Shouldn't there be questions, asked in a deep booming voice, like: Does your objective positively impact the world? Does your objective mitigate harm to others or the planet? Can you meet this objective with sense of moral rectitude? Can you meet this objective and still meet other private needs, like caring for your child or being a good friend or loving your spouse?

I think my idols may have actually considered these questions and unknowingly followed the way of Lao Tzu, who said: "In work, do what you enjoy. In family life, be completely present." Robert maintained close and witty correspondence with his sisters and parents, even when he was away. Peattie, too, seemed to have friends who enjoyed his cocktails. Their lives were in equilibrium in a way that our society doesn't promote when we discuss big-name achievers today, like Steve Jobs or Jeff Bezos. Kennicott and Peattie's lives were rich with friends and family, but both continued to work toward some sort of larger, amorphous goal. Their names aren't known around the globe, but didn't they still change the world? I ask myself still, did their lives have meaning? And, does mine?

THE OUTBACK

BETWEEN CHILDHOOD AND adolescence, social- and self-awareness cause a strange turbulence. The easy, wide-eyed connection that we had to the other—to mice or mountains or mastodons— begins to slip away, replaced by a type of high-alert anxiety. In adolescence, we begin to lose both fluidity of thought and focus, our minds filling with the minutiae of middle-school social status or honors classes and academic standing.

When I was thirteen, I remember sitting down at a long, gray lunch table in the cavernous cafeteria, filled with the smell of processed food and super-chilled milk. A few of my friends already sat at the table; others began to join. When almost all the places were filled, a friend who always sat with us came over to our table. She had a larger build and a mass of curly, dirty-blonde hair, and a saucy personality. For some reason, another girl with a long, dour face and straight brown hair said, "You can't sit with us." A couple other girls concurred.

My adolescent mind knew that an injustice was occurring. I remember simply countermanding the whole thing, saying, "Yes, she can." She sat down; the other girls were cold but not cruel. It was the last time I sat in the cafeteria that year. The next day, I asked a teacher if I could eat my lunch in the library, and she said yes. The rest of that semester, I ate happily and alone. I was always a fast eater, so when I finished up my lunch, I would walk up and down the long metal shelves of books and wait for something to catch my eye. One day, my eyes settled on two big words, "transcendental meditation." I took out the book and began to read. I took the book home and began to practice. Every evening before bed, I began to meditate.

Meditation allowed me to keep something that was slipping away—a sense of continuity, wholeness. It also allowed me to be more resilient in the face of teenage tumult; all those high-drama fights and the petty wrangling for position seemed removed from me. I ended up, I suspect, happier, healthier, and more confident because of meditation. I also ended up more connected. Meditation served as a touchpoint for me, a place where I could be myself and lose myself. Meditation was a kind of everyday magic.

As I got older, the stresses of a roiling and rippling marriage wore down my defenses. Meditation took a back seat to striving for marital harmony. It became a refuge that I entered less and less frequently, and inconstancy meant that the magic lost some of its power to revive, renew, and reconnect. A new type of magic, though, was entering my life—travel. Sometime in my late twenties, travel shifted from being a type of glorious adventure that I could ill-afford to more of a psychological necessity. Travel became a touchpoint on dull days and demoralizing days and dark days. Traveling became akin to entering a meditative state, a way to achieve true presence. It took a couple days, when a trip first started, to shed the tensions of trivialities and the stresses of strict schedules, but once shed, a miraculous realm opened up: the realm of childhood connection and preadolescent power; the realm of the open mind and the open heart; the realm of free expression and free will.

At twenty-seven, Australia was my freedom: wild animals, wide open spaces, and indiscriminate wandering. Connection came easy in the Land Down Under. With remnants of ancient Gondwanaland forests to the south and tropical rainforests to the north, plus more than 1,600 terrestrial vertebrate species, over 80 percent of which are endemic to Australia, traveling here was like unwrapping a present of biological wonder.

We were on a family adventure, me, my husband, and my parents, both in their late fifties. One of our first stops was the South Australia Museum in Adelaide, more than 850 miles west of Sydney. Its Megafauna Gallery housed a wonderful collection of extinct animals that had lived relatively recently. Life-sized dioramas placed enormous skeletons onto the red earth of Australia, and the backdrop was painted to show the flora with those same animals fleshed out and furred. We marveled at the enormous Diprotodon—a giant, extinct marsupial resembling a pouched mastodon—that roamed Australia ten thousand years ago. Its skeleton preserved huge buck teeth, flaring nostrils, ribs longer than my legs, and pigeon-toed feet. The drawing looked more like a big, sad dog with tiny ears and an elephant-seal nose. Even the museum artifacts had an outsized and otherworldly feel.

My strongest memory from this time feels more like a hallucination. We went to King's Canyon, a revered First Nations site two hundred miles from Uluru (Ayer's Rock) by car at the southern end of Australia's Northern Territory. King's Canyon is a sandstone wonderland, with sienna walls towering more than ninety meters high, a deep canyon spotted with scrubby, sage-green vegetation. A short, six-kilometer hike took us along the rim of the canyon, where we gawked at the cavernous views and beehive-shaped sandstone formations resembling the towers of Angkor Wat.

The land felt inhospitable. It was dry and vast, and even the plant life had haunting names like the Ghost Gum Tree, but still we were greeted by the chirruping of colorful Zebra Finches, with their intense red eyes, fluorescent orange bills and legs, and striking chestnut cheeks. Two Spinifex Pigeons sat nestled between pigeon-sized sandstone rocks, futilely trying to hide their brown-spiked crowns and their puffed-out tan breasts decorated with a thick band of white and a paintbrush-thin ring of black. A White-plumed Honeyeater, a petite lemon-yellow bird with a distinct brush of white feathers along the side of its neck, posed in a worn-out, broad-leaved shrub.

We met lizards sunning on the smooth, sandstone boulders. One lizard stood stoic and aloof, never deigning to turn his bulbous head our way. With his small body, marked with a white stripe so thick it looked painted on, and an overlong S-shaped tail, he had the air of a little Napoleon sitting on his throne on coronation day, his head dwarfed by an ermine collar and a gaudy red velvet cloak. Another lizard, painted burnt-yellow ochre and dabbed with light dots of lamp black, was stout-legged, pushing up his wide body like a bodybuilder, alert and ready to chase off a competitor.

We walked on for several kilometers, sweat dripping down our backs, our faces covered by fine netting to keep the flies away. The flies were thick. I remember looking down at my mother's slender, damp wrist. She had on a watch, with a thin strap, black bezel, and an iridescent purple crystal that shimmered in the light. At least a dozen small flies surrounded the watch, trying to push their bodies beneath the case body. Her watch looked alive, pulsing and wiggling.

Despite the flies and the heat, we were enjoying the solitude and expansive views. But perhaps the heat was getting to us. When we passed a bend in the path, we met the unexpected: a dark-haired model in all black, posing against the backdrop of the fiery canyon walls and the ultramarine sky. The photographer and gear must have dropped from the sky. The model appeared to be about to fly back into it, like a magnificent raven.

As we left King's Canyon, the sun had begun to set. We had been warned not to drive on the roads at night, but we had spent too much time hiking and still had a five-hour drive to make it back to Alice Springs. We began to drive down the road; the sky looked dusty, the landscape awash in a dull, dim orange-red. And then we saw it, a large white camel in the distance. We seemed to pass it in slow motion; the camel was towering, and luminescent, as if carved in Carrera marble by Michelangelo himself. Its dark eyes met mine; shivers went down my spine. Minutes later, a huge tawny-hued owl swept down in front of our car, stopped and landed on the pavement ahead, starred at us, and swooped back up into the sky. And then there were the kangaroos. We hadn't seen one all day. But now, with the sun nearly set and the air touched with coolness, the kangaroos were everywhere. As we drove, slow and steady, one hefty kangaroo hopped along with our car on the driver's side before jumping ahead and in front of the car to cross the road.

All through the long drive we were tense and on high alert, trying to avoid colliding with Australia's fine, beautiful beasts. We were also terrified by the high-speed road trains—semi trucks hauling multiple trailers. In the rural, wide-open Northern Territory, we were being overtaken by triple and quadruple road trains, pulling three or four trailers, stretching up to 53.5 meters (175.5 feet) long and cruising at speeds over 100 km/hr. These massive trucks are the feeding tubes of this rural area, ensuring that Australians living in the Northern Territory, a place nearly as big in area as Mongolia but housing, in 2008, fewer people than the island countries of Vanuatu or Barbados, are supplied with food and fuel.

Near the end of the white-knuckle drive, we pulled over in a parking area near a creek so the men could use the bathroom. The area lacked light and amenities, but it did allow for plenty of privacy. We all exited the rental car, stretching our tight muscles, staring into the night sky, searching for Ursa Major and Leo Minor and the Lynx. My husband had returned to the car, and my dad was within a few feet, when we heard a primeval, loud, low grunt and the sound of something large moving toward us. We panicked, and someone yelled, "go, go, go!" My parents jumped into the car. I remember pushing my husband into the rear seat, falling in after him, and slamming the door.

Lots of animals in the Northern Territory can kill a man—crocodiles,

Death Adders, Redback Spiders (an Australian version of the American Black Widow). Our creature was much bigger, but we never did find out what it was.

Days after our experience in the surreal south of the Northern Territory, we went north, far from the burnt sienna landscape into lush, green jungle. Even the drive into Kakadu National Park hinted at the type of lush fecundity that would make Annie Dillard's skin crawl: ten-foot-tall termite mounds, wallabies bounding about, and a wild dingo lurking in the distance.

By the time we arrived at a real billabong, a large, outstretched wetland backed by flat forests and a sheer line of cliffs, we were in a photographic frenzy, trying to digitally capture every lizard, each new tree frog, a crocodile, and at least eleven life-list bird species. We caught our breath, glimpsed lorikeets flitting through the paperbark trees in search of flowers, and decided to walk. We walked three kilometers, often through warm, knee-deep water, into a monsoon forest. In the distance we could hear rushing water, and we walked farther until we found its source: a river composed of beautiful pools, filling with water that rushed in through ancient boulders. We stood on the bank, mesmerized. And then we saw them—two great big, gray water monitor lizards sunning themselves.

My husband and I stumbled over the rocks to get a closer look, climbing atop a five-foot monolith to look down upon a monitor inching into the spray of the rumbling stream. The cool water proved just as irresistible to us. We slid into the water in our clothes—it was wonderful, cooling but still warm, sparkling and rushing with the strong current, so deep that I never touched the bottom.

We swam toward the monitors, and then, while we were plastered against a large boulder, one slid into the water and swam downstream. My husband stretched out his long arm, grazing the tail of the monitor lizard as it slid past, graceful and perceptive, soon to disappear beneath the bubbling surface.

Each day was a dream, filled with a magic that felt once in a lifetime, but might always be in reach, the magic of connection—connection to life, connection to nature, connection to all that is primal and shared among the creatures of the earth. Tree frogs sang us to sleep. The two-note calls of a family of four boobook owls woke us up. Flying foxes, bats as big as chickens, hung from trees as we rested at midday.

The biotic enchantment of Australia continued in Yungaburra, a town perched on the Atherton Tableland of Queensland, the north-easternmost state in Australia. Queensland itself is home to First Nations people—including Torres Strait Islander and Aboriginal peoples. It is also referred to as the Deep North, a reference to the Deep South of the United States, and the commonalities they share in terms of conservative politics and a deep-rooted history of racism. While the cultural diversity of the state causes tensions locally, the biological diversity drives a thriving tourist industry. Queensland ranks as Australia's most biodiverse state, supporting twenty-seven terrestrial and marine bioregions and more than one thousand different ecotypes. The majority of the country's resident mammals, birds, reptiles, amphibians, and plants call Queensland home.

One of the most unlikely animals to live in Queensland is the platypus, a twenty-inch-long mammal that seems to have stolen the most unusual features of other animals to call its own, sporting a duck-like bill, a beaver-like tail, and toxic stingers on the heels of its rear feet. This mishmashed assemblage of parts works. The platypus's duck-like bill is exquisitely sensitive, detecting electrical signals from the muscles of its favorite food, crustaceans, insect larvae, and worms buried in the muddy bottoms of rivers. The beaver-like tail provides stability while swimming, and those toxic stingers allow male platypuses to compete for mates. The stingers are found only on males, and the venom is most active during breeding season.

The platypuses draw visitors from around the world to the Tablelands, where more platypuses live than anywhere else, and we were among them. We woke early one gray morning and drove to a well-known viewing location: Peterson Creek, rock-filled and slow-moving.

It was misty and quiet; the thick vegetation seemed to hold in moisture and hold back sound. We waited and waited. A hint of sun could be seen over the horizon. A shadowy figure emerged, plump and surprisingly graceful, sliding over the glassy water. There it was, the elusive platypus.

The highlight, though, above all else, was the snakes.

Australia vaunts 172 snake species, a number of which have only been described by Western science in the last forty years, due to new discoveries and genetic analysis. Over half of Australia's snakes are venomous sea snakes (thirty-one species) and other elapids (eighty-two species), most of which likely evolved from a single species of Asian coral snake that arrived on the continent fifteen to twenty million years ago. About thirty of Australia's snakes are blind snakes, a group of fossorial snakes with vestigial eyes. Australia also houses nearly 50 percent of the world's python species, snakes that experts describe as more slender and less arboreal than we often imagine.

Australia is fortunate to have its own snake champion, Richard Shine, a professor of biology at Macquarie University, an emeritus professor at the University of Sydney, and the winner of a number of prestigious awards. Shine has said that "being interested in snakes is like supporting a football team that loses almost every game. You are part of a small but enthusiastic minority, while everyone else thinks you're crazy. You have only two options open: abandon the unpopular cause or try to persuade everyone else to re-examine their attitude." Shine chose the latter.

Shine has updated and elucidated the ecology of Australia's snakes, and his book *Australian Snakes: A Natural History* is a masterpiece. He has spent a lifetime shedding light on threats to snake persistence, including commercial exploitation for food and skins, bounties and roundups, pesticides and chemical contamination in the environment, and habitat loss.

One threat to Australian snake populations is an invasive species: the Cane Toad. The Cane Toad was introduced to Australia in 1935 to control the ravenous grubs of the Cane Beetle that were chomping away at northeastern Australia's sugarcane crop. The toads failed to control the Cane Beetle, but they did thrive in Australia. Today, more than two hundred million Cane Toads hop across thousands of square miles of Australia. This wouldn't be quite so bad if the Cane Toad were innocuous, but it isn't. Instead, Cane Toads are poisonous, with milky poison being particularly concentrated in the wide raised bumps of their parotoid glands. Cane Toads have been known to kill cats, dogs, and even people.

Cane Toads also kill Australian snakes. Shine and colleagues have described mammoth declines in the populations of the Red-bellied Black Snake, or Mulga Snake—a common and mildly venomous elapid with a shiny black back and bright-red sides—due to their low resistance to Cane Toad toxins. But all hope isn't quite lost. Shine also described adaptations among the Red-bellied Black Snakes that might save their populations yet: in some areas, the snakes are starting to shift their diet from Cane Toads and are developing increased resistance to their toxins. Although some snake species do learn to avoid toxic prey, like garter snakes in North America, the Red-bellied Black Snakes don't learn or adapt within their own lifetime; rather, the Cane Toads have acted as a selective pressure. Snakes that eat Cane Toads die and have fewer offspring. Snakes that avoid Cane Toads live and have more offspring. Avoidance and resistance presumably develop after many generations.

We saw our fair share of Cane Toads while exploring Australia, but we were looking for a different type of biotic thrill. So we went on to Lakes Barrine and Eacham, two magnificent water bodies located less than eight kilometers apart. Both lakes formed between nine thousand and seventeen thousand years ago, when water filled two maars, or broad, flat craters left behind by the explosion caused when hot magma comes into contact with groundwater. Maars are found throughout the world, reaching nearly five miles across in northwest Alaska and being particularly dense in the Eifel region of western Germany, which boasts seventy-five maars, both wet and dry. In the Tableland region of Australia, fifty-two different explosive centers are known, and two of them are protected by Crater Lakes National Park.

The blue jewels are surrounded by the verdant patches of the threatened rainforest, making the park a sapphire and emerald enchantment. We saw Water Dragons, admired the Saw-shelled Terrapins, and heard the cracking-whip call of the eponymous Whipbird. We touched magenta fungus, smooth and flat like smooshed Play-Doh, observed small gray birds with pert little tails, and chased quick-moving skinks. We were like addicts, constantly craving more. More life. More fecundity. More connection. We walked and walked, circumnavigating the lakes.

Finally, we saw it, a shimmering gray-green glint reflecting from the back of a Common Tree Snake, only a couple feet long. It moved like mercury, smooth and effortless. Its dorsal scales were large, each one lined on three sides in black, and one side in light teal. The belly gleamed lime green, then sky blue. And then it turned and headed toward us, seeming to pin its oversized, round eyes on mine. A shiver raced through me. I stood open-mouthed and awed, until she moved away with that liquid flow.

Common Tree Snakes are harmless, to us at least. They rarely bite, and their mildly toxic venom only seems to affect prey that is already being swallowed. They occupy a variety of habitats, wet forests, dry woodlands, and even suburban gardens, all in search of their favorite prey, frogs and skinks and Asian house geckos. The easy meal of an Asian house gecko, which sticks to the sides of verandahs and sheds, also means that the Common Tree Snake is one of the most frequently seen snakes in Queensland. It might be common, but that made it no less impressive. In many ways, watching the Common Tree Snake was like watching a puma, all grace and speed and danger.

Our next discovery was more like sitting with your soft, lazy, old Golden Retriever. We walked through the woods, continuing to skirt the edge of the lake. At one point, we came to a small clearing in the dense growth; a two-foot-diameter tree had fallen. I was moving closer to the tree, edging to see the water more clearly, when I saw it. A large snake, as thick as a man's arm, nestled among the nine-inch-long dead leaves of the fallen tree. It was an Amethystine Python.

Amethystine, or Scrub, Pythons are Australia's longest, stretching more than sixteen feet. Pythons are a family of large ambush predators, known for their camouflaged patience, at least until they strike at passing prey. Pythons rarely strike at people, and for the most part they are no threat to us, except if you leave your teddy bears out. One midwinter's day in northern Queensland, an Amethystine Python was caught chowing down on a little girl's fluffy white-and-blue stuffed-animal cow. By the time snake catchers arrived, the six-foot python had a bulge twice its normal diameter in the middle of its body. Veterinarians were able to surgically remove the stuffed animal, and the snake was handed over to rehabilitators who planned to release it back into the wild.

When not chowing down on plush toys, Amethystine Pythons can typically be found preying upon birds, bats, and opossums. Large individuals have even been observed eating Brown Bandicoots and Bennett's Tree Kangaroos, cryptic marsupials weighing up to thirty pounds.

Amethystine Pythons, named after the creamy iridescence of their scales, which shine pale purple in some light, have been known to Western science since at least 1801, when they were described from a now lost type specimen by Johann Gottlob Theaenus Schneider, a German natural historian who did his best work from behind a desk. While known as being the longest Australian snake, occupying the far northeastern corner of Queensland, Amethystine Pythons also inhabit nearby Papua New Guinea.

Only more than two hundred years after the species was described were researchers able to conduct the first detailed field study of Amethystine Pythons, opening our eyes to the ophidian giant's behavior in the wild. Negotiating dangerous gorges and perusing roadsides, the researchers found that female Amethystine Pythons were more active and obvious in the late afternoon than their male counterparts. Other research supports this. Female snakes might need to "maintain high and constant temperatures" to expedite successful reproduction. The researchers also observed something new: reproductive congregations of Amethystine Pythons in the dry season, facilitating mating and pre-coital combat between large males.

We had arrived in Queensland during the wet season, so rather than mating and fighting, the Amethystine Pythons that we spied looked heavy and sleepy, warming up in the early-afternoon sunshine. We were able to edge close to one python, wrapped tight like a Solomon knot, its chin resting heavily on its back. The large snake was so still that we could stand and stare, taking in the details: its tan base color, the band of two dark-brown strands interwoven like the simple guilloche patterns rimming ancient Roman mosaics, the purple shimmer of the upper body, the green sheen of the small, compact scales near the end of the tail, the cream of the supralabial scales making it appear as if the snake had just finished off a vanilla ice cream cone, and the steady stare of the elliptical pupil surrounded by a soft mauve iris. In that quiet, meditative moment, the magic of connection was working. It had offered something that I hadn't felt in a long time. Peace.

PERU

 IN 1998, A TWENTY-one-year-old American and his father took a trip to Peru. What started out as a typical, organized tour up the tributaries of the Amazon soon became a herping adventure that included a bona fide herpetologist from a Texas zoo.

The tour started in Nauta, a common launching point to explore the headwaters of the Amazon, particularly the Nauta Caño, Nahuapa, and Tigre Rivers. Nauta was well-populated, with buildings in all shapes and sizes: small, cement-block structures plastered and painted bright cerulean, sea green, and sky blue; wooden homes with elevated front porches adapted to the frequent floods of the river; a sprawling turquoise clapboard building with a corrugated metal roof and painted-on advertisements for the mayoral candidate—"Let's go, neighbor [Vamos vecino], for Beto Saavedra." The town's dock was a steep, muddy bank with twenty long, squared-off logs secured in place by thick wooden stakes. Boats of all kinds came up to the dock, ones with long thatched roofs, canoes filled with fish, and speedboats too.

When the young man and the group of eleven other people on the riverboat tour finally launched, they were captivated by shoreline birds: groups of two dozen Great Egrets with their majestically curved white necks and strong yellow bills; a pair of Jacanas, dark-brown birds with red wattled foreheads; and a Horned Screamer, heavy-bodied and known for its echoing scream, perched on a branch that seemed much too slender to support the bird's nearly eight-pound heft.

As the cruise progressed, the boat stopped at isolated preserves along the riverbank. The group would take short hikes, with the herpetologist pointing out iguanas stretched along a branch high up in the trees and finding a little stump sheltering a pair of Fringed Leaf Frogs that had laid eggs. The snakes were impressive too, snail eaters and tree boas. Plus, all that herpetological splendor was tucked in among the spiky yellow and red birds of paradise, broadleaved and thick-veined Iodine Plants, and Hot Lips Plants with tiny white flowers protected by conspicuous red bracts, like two wide, cherry-stained lips. Termite mounds, bullet ants, and leeches, perched on low vegetation, completed the picture.

At night, most passengers were safely ensconced on the boat, reading and enjoying a cold beverage, but the hard-core were out again, with high rubber boots and flashlights, on the prowl for more herps. They weren't disappointed. There was a Smoky Jungle Frog, robust and dark, with a strikingly marked face, a black stripe extending from the nostril, through the eye, above and beyond the ear, ending in a broken line, as if marked by a brush running out of paint. These frogs make their displeasure known when grabbed, emitting an oddly high-pitched screech and mild toxins that irritate the skin. Three species of caiman also found their way into the hands of the herpetologist and the young man. There was the Spectacled Caiman, named for the ridge connecting its eyes that looks like the bridge of eyeglass frames. The most widespread in the Americas, these caimans have a complex form of acoustic communication including hatching calls that encourage mom to open up the nest when the young have begun to push out of their eggs. Then the group found a Schneider's Smooth-fronted Caiman, the second-smallest crocodilian in the world, which hides in underwater burrows during the day. Finally, they grabbed a Black Caiman, a large-bodied apex predator of the Amazon that feeds on just about anything and occasionally bites people.

After the river cruise was over, on the last night of the trip, the young man and the herpetologist drove fifteen miles outside of Iquitos, road cruising for snakes. After finding a few dead on the road, they spotted a three-foot-long snake along the side bordering a secondary-growth forest. The snake was beautiful, patterned with transverse bands of red, then black, then yellow, and black and red again. It was also about to slide off into the dense forest. The young man and the herpetologist hopped out of the car quickly; the herpetologist thought from afar that it might be a *Lampropeltis,* or milksnake species. The young man, impulsive and unwilling to let his quarry get away as it began to move, suddenly lunged forward.

Let me provide you with context. The young man was an avid amateur herpetologist, having spent his boyhood catching snakes in rural Tennessee. One of his first snake memories, from his kindergarten years, was of his dad catching a garter snake in the backyard garden. The family kept the snake for a week before releasing it, and the experience stayed with him. One snake and he was hooked. Soon he was out in the chicken coop catching Rat Snakes or chasing after racers sunning themselves in the grass or lifting the catchment for the gutter's downspout to find Black Kingsnakes.

When the young man was in first grade, his dad carved a pond out of the rich Tennessee soil with a bulldozer. Now he'd go out at night to the pond, catching watersnakes and a single Ringneck Snake. The woods yielded more Ringneck Snakes and Red-bellied Snakes and Rough Green Snakes. Near the woodland edges, he'd catch hognose snakes and more rarely a Prairie Kingsnake or Pine Snake. When he was older, he began going farther afield, wandering across the rural properties adjoining that of his family and passing time going up and down shallow, stony creeks looking for fierce Midland Watersnakes and gentle Queen Snakes.

Thus, by the time he was twenty, the young man had honed the instinct for just grabbing any snake he saw, unless it was a Copperhead. And here in Peru, that instinct kicked in.

The young man lunged, grabbing the colorful snake with one hand. The snake quickly turned around and bit the fleshy part of his thumb four times in rapid succession. He immediately knew he was in trouble, for two reasons. First, he saw the way the snake forcefully chewed his hand. Second, it hurt, instantaneously. Venom was working its way into his body. It turns out that the maybe milksnake was actually a South American Coral Snake, *Micrurus lemniscatus*. It's worth mentioning here that the old rule, red touches black, friend of Jack; red touches yellow, dangerous fellow, doesn't work south of Mexico City. In this case, red touched black, but it was no friend of Jack. The herpetologist identified it definitively.

The young man and the herpetologist were thirty minutes from Iquitos, a city of 250,000 people in 1998, filled with zipping scooters and three-wheeled taxis. The city buses were stuffed with people. The pavement simply stopped at the edge of town. At this time, Iquitos was isolated in many respects, the farthest inland Amazonian port, only accessible by river or air. When they pulled up to the hospital, it was surrounded by a cement wall embedded with large panels of chain-link fence. It allowed passersby to see the small complex of whitewashed, hip-roofed, one-story buildings within. They went inside the small hospital, equipped with a dedicated staff relying on glass syringes. Geckos poked their heads into the windows and came and went at will.

The young man was in pain as an excruciating burning sensation slowly worked its way up his arm. He was admitted to the small, poorly equipped ICU. The pain had reached his shoulder. The nurses began to draw blood and administer fluids, but the sequence of events started to get blurry here. The pain in his arm was now tremendous as venom broke down his nerves. It was so painful that it was hard to focus. The nurse gave him opioids, which did nothing for the pain. They also gave him a couple vials of antivenom, all they had available, but being a bigger guy, six feet one-half inch and nearly two hundred pounds, the sort who boxed extracurricularly, he needed ten to twenty vials.

By the next morning, after the nerves had been damaged as much as they could be, the pain went away, along with all sensation. That's when the paralysis kicked in. The paralysis was generalized, but it seemed to affect his legs more than his arms. He began to have double vision because the muscles around his eyes were frozen. He was being watched around the clock. He was having trouble breathing, and there were no ventilators at the hospital. If his breathing got any worse, he would have to be flown to Lima.

He hadn't slept for two straight nights. After a few days, he was able to sleep, sort of. His dreams were red-hued, bright and vivid. He felt scared, with a sense that evil was all around him. But slowly, and after many nights, the wretched dreams became less terrifying. He made friends with the geckos on the wall.

He couldn't chew. Only after a few days did he finally begin to eat mushy foods, but swallowing was exceedingly difficult. The effects of paralysis wore off gradually. It took seven months to regain feeling in his body. By day four or five, he was so hungry that he found himself thinking he might die of starvation as a young man might, and though he could barely move, he willed himself upright to painstakingly stagger to the nurses' station to plead for food. He was fifteen pounds lighter by the time they released him from the hospital in a wheelchair on day seven.

You'd think that after a traumatic experience like this, one might be turned off by herpetology, but that wasn't the case for the young man. Four years later, we met while taking that tropical herpetology course in Nicaragua. In fact, the young man, now my husband, became well-versed in venoms and their effects.

Nearly twenty-five years later, he can still rattle off the types of toxins and what they do. For a coral snake bite, one culprit contained in the venom is phospholipase A2 (PLA2), an enzyme that human bodies produce for digestion but that does severe damage when injected into the body. It breaks down the fatty myelin tissue around the nerves, deteriorating them enough to make people lose feeling completely. However, the cause of paralysis is twofold. Phospholipase A2 also decreases acetylcholine, a neurotransmitter needed for communication between the brain and muscles, while the nonenzymatic proteins in venom block the acetylcholine receptors.

As difficult as it is to accept, some snakes can kill us. Viperids—those fanged vipers, some with deep heat-sensing pits at the front of their wide faces, some with rattles (for example, rattlesnakes) and some without (for example, cottonmouths)—can kill us. Elapids can kill us, too, from multicolored wonders like coral snakes to dark mambas to mesmerizing hooded cobras. Atractaspids, a family of snakes found in the Middle East and Africa, are also deadly, although most of the snakes are too small to envenomate anyone easily. Even some species in typically harmless groups of snakes, like colubrids, can kill us. The Boomslang, a gorgeous turquoise-and-black denizen of sub-Saharan Africa, has a huge gape and particularly potent venom. In fact, Dr. Karl Schmidt, a herpetologist from Chicagoland, documented his symptoms after being bitten by a juvenile Boomslang in his lab at the Field Museum.

According to Dr. Schmidt's own account, he had absentmindedly grasped the snake, which was being passed to him by a colleague, in a way that allowed one fang and other teeth to enter his left thumb. The punctures bled profusely. In the early evening, Dr. Schmidt took the train to Homewood, a suburb located about twenty-five miles south of the city.

During his trip, he became nauseous. Soon, he developed a fever, feeling deeply chilled and shaking all over. By 5:30 p.m. he began to bleed from his gums, but it didn't stop him from eating milk toast around 8:30 p.m. and heading to bed at 9:00 p.m. He slept for three hours and twenty minutes before waking up to pee; his urine was mostly blood. He suffered from abdominal pain and vomited at

4:30 a.m. and then slept another two hours. When he awoke, his fever was gone, and he drank his last cup of coffee and ate his last breakfast: cereal, a poached egg on toast, and applesauce.

His urine and stool were mostly blood, but he felt better, optimistic even, and called the Field Museum to tell them that he'd be at work the next day. Unfortunately, this wasn't to be. By noon, his breathing became labored, and by 3:00 p.m., Dr. Schmidt was pronounced dead from respiratory failure. An autopsy found massive internal hemorrhages in the small intestine and brain, with smaller hemorrhages throughout his body.

While Dr. Schmidt's life was cut short, he contributed greatly to the field of herpetology. As the chief curator of zoology of the Field Museum, he massively increased the size of the herpetology collection. He also conducted fieldwork in South and Central America, wrote more than 160 books and papers on herpetology, and named more than two hundred species and subspecies of herpetofauna.

As Karl Schmidt could attest, snakes can kill us in horrible, disparate ways, depending on the chemistry of their venom. The venom of the Boomslang is considered proteolytic and hemotoxic. This means that Boomslang venom contains enzymes that accelerate the breakdown of proteins and destroy red blood cells. Hemotoxic venoms, which many vipers produce, affect the heart and cardiovascular system. Most people who die from these venoms simply bleed out. Other venoms are neurotoxins, which disrupt nerve impulses and cause paralysis. Most elapids, including mambas and kraits and coral snakes, have venom of this type. Cytotoxic venoms simply affect the tissues around the bite itself.

But snake venoms aren't all bad. Researchers in Australia, Singapore, and Switzerland have been working together to investigate new drugs and diagnostic tools born from the special properties of Eastern Coral Snake venom. In fact, researchers have recently found that Eastern Coral Snake venom can form a strange eight-molecule complex that looks like a crown.

The benefit of this complicated form is that the venom can target more protein receptors than a single molecule. For example, the neurotoxin targets protein receptors in mammals that control nerve and muscle cell communication. For the Eastern Coral Snake, this probably means being able to immobilize predators more effectively, but the venom might benefit people too. Researchers are now investigating the use of Eastern Coral Snake venom as a molecular probe to help diagnose Alzheimer's and Parkinson's disease.

Snake venoms have already resulted in other medicines. Brazilian Pit Viper venom has been used to make a blood pressure medication, captopril. Jenny Bryan, a writer for the *Pharmaceutical Journal,* relayed the following:

> In the early 1980s, hypertension conferences were routinely enlivened by the poisonous Brazilian viper, Bothrops jararaca. With its striking zig-zag markings and aggressively protruding tongue, images of the snake were a welcome break from graphs and tables in presentations about captopril—the first of the angiotensin-converting enzyme (ACE) inhibitors, whose effects on blood pressure mechanisms mimicked those of the snake's venom. When the cardiovascular juggernaut alighted in Sao Paulo, Brazil, for a major congress in 1984, there was even an opportunity for delegates to visit a snake farm and see the beast in all its glory.

Today, captopril isn't all that popular, but the venom of the Brazilian Pit Viper launched a revolution in blood pressure treatment that presaged today's cocktails of ACE inhibitors, beta blockers, diuretics, and exercise.

Still, some snakes are deadly, and movies and literature know it well.

Movies like *Anaconda* or *Snakes on a Plane* use our fear of death to startle and entertain us. These movies exaggerate and distort the real behavior of snakes. In *Snakes on a Plane*, snakes slither around, disrupt electrical equipment, and lunge at ankles with predetermined precision, often striking with a loud hiss. The loud hiss is particularly annoying, since in real life that hiss would give away the snake's position and intention, causing prey to flee, and because in real life, I've never heard a snake make any noise at all as it bit or fed. Snakes attack in the bathroom, snakes sit in oxygen compartments awaiting the moment they can launch at people's faces in synchrony, and impossibly large snakes, weirdly weightless, await unnoticed in the throw-up bag. No movie does a better job of showing the extent of human's irrational fear of snakes than this.

Literature isn't much better. Stephen Crane, an innovative American poet, novelist, and journalist, whose novella *The Monster* remains controversial for its racist overtones and who died of tuberculosis in a German sanatorium at age twenty-eight, describes a scene where a man, while walking his dog, hears the rattle of a snake.

> Suddenly from some unknown and yet near place in advance there came a dry shrill whistling rattle that smote motion instantly in the limbs of the man and the dog. Like the fingers of a sudden death, this sound seemed to touch the man at the nape of the neck, at the top of the spine, and change him, as swift as thought, to a statue of listening horror, surprise, rage. The dog, too—the same icy hand was laid upon him and he stood crouched and quivering, his jaw drooping, the froth of terror upon his lips, the light of hatred in his eyes.

The story emphasizes the quick association made between snakes and death, but as is most often the case with snakes, despite the man's fear, it isn't his death we await. The story ends with the man beating the snake to death with a stick and turning to his dog with a wide grin saying, "Well, Rover . . . we'll carry Mr. Snake home to show the girls."

Some stories at least try to proffer explanations of how snakes came to be. One African American folktale relates the origin of the rattlesnake's venom and rattle. In the tale, when God first made snakes, they were beautiful and harmless. They couldn't fly or run; they were just down in the dirt all day. And because they were on the ground, the other animals kept stepping on them. The situation became so bad that one brave snake begged God for help. God gave the snake venom for protection. The snake began to bite any animal that came near, and the venom worked well, so well that the other animals called a congress to complain. God heard the other animals' complaints and gifted the snake with rattles, saying: "When you hear something, shake your tail. That'll be a warning. If it's your friend, he'll stop and pass the time of day with you. And if it's your enemy, he'll just keep coming, and after that, it's you and him."

This folktale approaches a fundamental truth about people and snakes. We are afraid of them, but they have more reason to be afraid of us. I suspect that more snakes die each year on my father-in-law's farm in Tennessee than people do from snakes in the whole country (about five) during that same time period.

In fact, my husband's own experience with being bitten by a coral snake highlights a key characteristic of many fatal snake bites, especially in the United States: people were messing around with venomous snakes. In the last decade, people have died from venomous snakes in the United States as a result of bites during Pentecostal religious rites in Kentucky, kind attempts to move a rattlesnake off the road or out of a garage, a suicide attempt involving a Monocled Cobra, and attempts to kill a rattlesnake. Although some tales are legitimately tragic, most deaths in the United States are avoidable. This isn't the case everywhere. In the Global South, venomous snakes represent a serious health issue exacerbated by the scarcity of antivenom and delays in administering it, but in the United States the situation is different.

The most common reaction to venomous snakes is fear, pure and simple, but some people have gone to great lengths to master their vulnerability. Bill Haast, born in New Jersey in 1910, developed a keen interest in snakes as a Boy Scout. By the age of thirteen, Haast had already been bitten by a Timber Rattlesnake and a Copperhead. Soon, he was learning how to handle venomous snakes in earnest, and by age fifteen, he was extracting venom.

Bill dreamed of starting a snake farm, but he needed money first. After a stint working at a speakeasy, Haast also needed a way to support his first wife and the baby they were expecting. He began to study aviation and soon moved to Miami, where he worked as a flight engineer on Pan Am airlines, flying around the globe.

By 1946, Haast purchased land south of Miami and started constructing a Serpentarium while still working for Pan Am. Those Pan Am flights allowed Haast to bring home exotic snakes from around the world. It was then that Haast started extracting venom from snakes in earnest. By 1965, Haast was extracting venom up to one hundred times a day from sixty species of snakes. The venom was used for medical research. Haast was also practicing mithridatism, injecting himself daily with nonlethal mixtures of thirty-two different venoms to build up immunity. It seemed to work—he became immune to elapid venom, although not to the venom of pit vipers. Haast was bitten more than 170 times by venomous snakes, including cobras, green mambas, and a Common Krait. He survived them all. Plus, he donated his blood more than twenty times to save the lives of people bitten by snakes for which no antivenom was available. Haast lived to be one hundred, dying in 2011.

For my husband, being bitten by a coral snake didn't change his life or how he felt about snakes. He remained fascinated by them. He still plans trips around certain snake species that he would like to see, and only with age did he become more cautious about grabbing them. In our family, we tend not to grab snakes when we see them, instead greeting them and showing them respect by letting them be.

Moreover, Mark was keen to go back to Peru, and we did, twenty years after his incident with the coral snake.

Peru is like two worlds to a tourist—one a world of mysterious, ancient mountains, and the other a world of powerful rivers and steamy jungle. There's the world of alpacas and Machu Picchu and the world of floating towns and motokar rides. My husband was ready to return to the jungle, and so we did.

We flew from the airport in Miami. Its carpet of black and gray polyester, interwoven with red and white threads, created an infinity of boxes that seemed to entrap the hopes and dreams of travelers until releasing them onto their planes. I remember watching a man who was sitting with one leg crossed over the other, head up to the ceiling, impatiently glancing at the large chrome watch on his thick wrist. A girl, holding up her phone, filmed herself sticking out her thick, pink tongue and then smiling toothily. My father-in-law waited with us, too, about to return to the place where his son nearly died. He was looking through pictures, on a giant iPad, of the Tennessee farm he had left only a few hours before. These were people in stasis, people whose intersecting dreams were on hold until they were released from airport purgatory.

But purgatory has a purpose. It allows us to expiate our sins. It cleanses us and purifies us and prepares us for the journey ahead. After airport purgatory, any destination is heaven. Our heaven was a brief stop in the bright-light city of Lima and then the final touchdown in Iquitos. From Iquitos we took a van to Nauta, about one hundred miles south, to get on a boat along the cecropia-lined north bank of a major tributary of the Amazon, the Río Marañon. The boat was called the *Clavero,* and it stretched ninety feet long and sixteen feet wide, holding eight passengers and nearly as many crew. Built in Paris in 1878, it protected the river from foreign encroachment, transported people and goods for expeditions on the Río Purus in 1905, and offered mail service until 1933. When we boarded the restored vessel, we were introduced to our captain, Dante. Clearly, we had the right guide for this airport afterlife.

We explored the headwaters of the Amazon, taking in Peru's remarkable biodiversity, a biodiversity that extends even to ants. One sample of ants from Peru's Manu National Park yielded "the most species-rich point sample of a canopy ant fauna ever documented."

Where we were, the nests of ants dripped from trees, like curtains of stalactites. Sloths sat high upon branches with iguanas and Saddleback Tamarins nearby. We spent hours on the deck of the boat staring at birds with all sorts of exaggerated features, like the giant, multicolored bill of the Aracari and the spine-like structure that grows from the skull of the Horned Screamer and the long, bright tails of Oropendolas. We spent days like this, feeling E. O. Wilson's biophilia, stunned and stuttering with each toucan that flew across a sky dotted with bright, lit-from-within clouds and every approaching canoe nearly brimful with armored catfish that would nourish several families.

During our first hike, at Estación Shiringal along the Río Samiria, we watched leaf cutter ants carry rolled bits of leaf across the forest floor and latex drip from wounded trees. Millipedes marched around, armored and prepared for battle, and an Amazon Sheep Frog, with a pointed nose and oakleaf motif down its spine, darted out of the way of our footfall. It took us hours to go a half mile, with each step full of life—diminutive snails and parasol-like fungi and a big eared fishing bat clutching a hanging vine.

At one point, we went downriver a bit more and entered a disturbed forest, with clear signs of people. The forest was oddly quiet, and a Common Squirrel Monkey watched us from afar. We soon came across something that made my husband's heart ache: an Aquatic Coral Snake (*Micrurus surinamensis*), mostly black with a pattern of two thin bands of yellow alternating with a broader band of red, each colored scale dotted with a hint of black. The coral snake's head had been cut off with a machete not long before we arrived.

This was a sign that the journey back to Peru wasn't complete. A few days later, while in Iquitos, we decided to visit the hospital where the careful attention of nurses had saved my husband's life. We walked in, and I explained in my slow Spanish that my husband had been bitten by a coral snake twenty years ago and had stayed there. Someone ducked into a door behind the reception area; someone else came out. It was Mercedes, a nurse who had stayed with my husband, feeding him, singing to him as he had recovered from his ordeal. La Clínica Adventista Ana Stahl had changed in twenty years, but not much. The front was newly painted in dark green, and the interior was bright and modern with no geckos crawling on the walls now, but the people, with their warm hearts and big smiles, were still the same.

GUMBY

 MY FIRST DAY OF work, I marched up to a log cabin. The door was hard to open; I pushed and pulled. Finally, I figured out the latch and walked in. The odor of mold and crayons permeated the dark space. The cabin was small, with four tight rooms. Two were the size of large bathrooms—those were the director's office and a preschool room. Two were larger; that's where the kids played when the weather was bad. It also was where the camp's animals lived—rabbits, snakes, lizards, and turtles at the ready to show the campers.

When I walked into the cabin, I met Sarah. She would train me. Sarah was in her early thirties, her hair long, dark blond, and tightly waved, her clothes loose, baggy, and secondhand. She smelled like sandalwood. The summer camp was all about connection to nature, and Sarah practiced what she preached. She left a light footprint on the earth, and that meant buying things used, foraging for food, and making her own teas and herbal remedies. Sarah spoke slowly and carefully. She thought deeply about the way she lived and the way that she guided the children. She was a wonderful teacher, but for me, working at the summer camp hadn't been the plan.

I was twenty-seven. I had finished writing my dissertation and defended it the previous winter. I landed my dream job, as a naturalist, in a three-thousand-acre park in Nashville. My husband and I packed up half of our stuff into two SUVs, one dog in each, and drove eight hours west.

But when we arrived in Nashville, it felt wrong. My stomach was upset, and my sixth sense screamed no at every house we visited with the real estate agent. A few days later, Mark went back to North Carolina to get the rest of our belongings. I moved in temporarily with a newly divorced woman. She cried a lot the first night I was there, finding little solace in her farm and horses. I slept on the floor of a room with new, thickly padded cream carpeting. At least my comforting dog was adjusting well; she made friends with the cat.

I spent a lot time thinking. My stomach still roiled. This didn't feel like the right situation for us. And then I found out I was pregnant. I called Mark and said I couldn't be in Nashville. Mark agreed. We brought our carloads back to North Carolina.

After this change of plans, I needed to find a part-time summer job. I decided to work with kids at a nature-based summer camp at Leigh Farm, an antebellum complex-cum-city park. The farm represents Durham's complex and overlapping histories, including a legacy of displacement of Tutelo- and Saponi-speaking Indigenous people by European colonizers, the practice of racialized slavery, and a movement to protect a little land for both people and wildlife in a rapidly urbanizing area.

After getting acquainted with Sarah and learning more about what I'd be doing at the camp, someone walked effortlessly through the big, wooden cabin door that I had so much trouble opening myself. He strolled into the dim room. What I noticed first was the tattoo on his right hand, a black, red, and yellow snake interlaced through his fingers and moving up over his wrist. It was a harmless Scarlet Kingsnake, but most people would have thought it was a venomous coral snake.

The man was about Sarah's age. He wore a tight T-shirt, long shorts, a red cap, and no shoes. He wasn't particularly tall, but he stood erect like someone who had never spent days on end hunched over a computer or cramming for finals. His shoulders were wide, and his chest barreled out; he was lean but powerfully built, like a wolf. His light-brown hair was cut close to his head; his blue eyes flashed and flickered. His name was gumby.

My first instinct was to be wary of gumby. He was confident, maybe even arrogant, and he seemed like someone who wouldn't hesitate to call you out on mistakes. It wouldn't matter if your error was innocent or bullshit. He exuded this with his erect posture, his unflinchingly direct gaze, and the hard line of his mouth.

Eventually, the three of us got to talking about the natural world, about the rhythms of nature, and about the wildlife of the Piedmont. gumby brought up snakes, and I perked up. He said that snakes can't dig holes. I perked up even more. I took a deep breath and said something like, well, that's not quite true. Some snakes can dig holes or burrows. They're not digging those little holes that everyone calls a snake hole, but some species can dig. I could tell that gumby didn't believe me. He didn't come out and say it, but his face changed.

The next day, I brought gumby the best evidence I could muster on short notice: a journal article about Bullsnakes in lab conditions that dug burrows. The abstract said it all:

The Bullsnake (*Pituophis melanoleucus sayi*) exhibits head morphology and stereotyped motor patterns adapted to excavating. Spading actions by the snout are followed by scooping a load of sand in a head-neck flexure, moving posterior, and dumping the load away from the excavation at varying distances. An excavating Bullsnake can move as much as 3400 cm^3/h. This behavior was seen in adults of both sexes and in a juvenile and may function for obtaining food (especially pocket gophers), digging a nest or a retreat. Bullsnakes respond positively to the soil from pocket gopher mounds.

The author of this gem, Dr. Charles Carpenter, was a distinguished herpetologist for whom a lizard, *Anolis carpenteri*, is now named. He spent at least thirteen hours of his productive life watching footage of eight Bullsnakes. Of those thirteen hours, the Bullsnakes used just two to dig. My favorite part of this article is a beautiful hand-drawn diagram sequencing the excavation technique of the Bullsnake, starting with spading, moving to scooping, then backward scooping (that is, "carrying a load of sand"), and straightening to drop the load. I'm pretty sure the diagram cinched it for gumby.

By then, I knew Bullsnakes well. I had taken care of them back in Illinois, before I had started college. I had even seen a Bullsnake in the wild, in a place few people ever go to see anything at all: North Dakota. On a lovely day in June, I spotted a Bullsnake sunning next to a hole on a high hill with short grass. It was beautiful.

Bullsnakes are largely diurnal and are one of a handful of snake species known to actually make their own burrows using the sinusoidal movements of their strong necks. These burrows can be used for escaping devastating prairie fires or laying eggs, although some females will lay their eggs at a communal nesting site, typically in June or July.

In many ways, the Bullsnake is the quintessential snake of the North American Great Plains, and a full 25 percent of its former range was tallgrass prairie, but today Bullsnakes are typically associated with sand prairies and old fields. Bullsnake populations have recently declined in the northern part of their range, North Dakota included, with habitat loss being a major threat to their populations.

After the Bullsnake exchange, gumby developed a modicum of respect for me. Not much though—that still needed to be earned. gumby and Sarah lived in the old Leigh farmhouse that sat on the property. It didn't have heat or air-conditioning, and gumby didn't care. gumby probably belongs in a different century; that's part of his allure.

The farmhouse, covered in wide off-white German siding, also served as the lunch spot for the summer camp. The house had a strange layout. It was one and a half stories tall, supported by a stone foundation. Two stone chimneys framed what was once the back door. An L-shaped porch edged the side of the main part of the house and covered the back of an 1880s dining room-cum-kitchen addition. We'd sit on the porch with the kids and then watch them explore. gumby could often be found ensconced on the porch steps as if it were a throne, with kids running up to him asking for the next nature challenge.

That summer was hot and humid. I was pregnant and sweaty and spent my days taking kids on quiet nature walks, facilitating their silent observation of nature with sit spots, teaching kids how to take crayon rubbings of the cemetery stones, carrying kids out of mud and muck, and finding box turtles. One searing afternoon, my body had enough. gumby said I didn't look right and brought me into the farmhouse. He plopped me down in front of a whirling metal fan. The house was creepy, and I was a little scared to be alone. That didn't stop me, though, from closing my eyes and passing out for an hour.

I was right to be scared. Years later, gumby and Sarah told me a story: One evening they returned home after eating dinner at a nearby café. They stepped onto the porch and went to the breezeway door. When gumby put his key into the keyhole, he and Sarah each felt a wave of dread, but neither said anything to the other. gumby felt like someone close to him had died or gotten badly injured. Sarah felt a presence looming over her, and the dread felt more like terror. The feeling that someone had died was so strong that gumby went to the phone to call his family. When he picked up the phone, he heard the dial tone for a second, but then the phone cut out completely. He couldn't make a call. Sarah went up the groaning, narrow staircase to her room. Soon, though, she came back down making a beeline out the door; gumby was heading that way too. That's when they acknowledged to each other that something wasn't right. They spent the next few hours with Sarah's parents and called the telephone company the next day; of course, nothing was wrong with the telephone line.

My own experience in the house wasn't nearly so spooky. I woke up groggy and weak but exceedingly thankful that gumby had brought me to the fan. He had watched out for me and the little human growing inside me.

Eventually that magic summer ended. My belly was big, and months later I gave birth to an incredible little boy. I had hoped to stay home with that sweet boy for five years before working again. Life didn't work out that way. Two years later, I earned a competitive fellowship with Duke University's Thompson Writing Program. They were looking for scientists who were interested in teaching and teaching about writing. I was in.

A couple of weeks into my second semester teaching, I contacted my friend gumby. Over the years, my sense of gumby had shifted. He came to seem like a cross between curmudgeonly, sarcastic ecoterrorist Edward Abbey and wandering, roaming Jack Kerouac. Still, after over three years of friendship, I didn't even know gumby's real name. He went by gumby, he didn't capitalize it, and he didn't like it if anyone else did either.

I asked gumby to speak to a class of master's students I was teaching the next fall. gumby refused out of principle; I'd thought he might. He said, "I think we need more passionate scoundrels, trespassers, and hobos out there in the fields, not more clean-kneed number crunchers from the cracker factory."

I also learned that gumby had been dealt a recent blow—he was heartbroken by a spirited, black-eyed whippersnapper who had been "corrupted" by higher education. gumby was depressed, living in a tent (houseless, but not homeless), and refused to teach environmental education to adults anymore because they were irredeemable. gumby's words touched a lot of nerves. He was struggling with things I was struggling with. He was asking questions I had been asking. We had found different answers.

gumby and I continued to correspond. He wrote me a platonic declaration of love and respect. He recognized the "sweet, youthful soul" that lived in me and declared, "Thank God Duke has you! I'm sure you're infusing such light and wonder and passion into your students, and I'm glad just knowing that you're out there somewhere sharing the beauty that's uniquely YOU!" I cried when I read his note. I cried because his independent soul had been trampled and because mine had been too. I cried because he recognized something in me that I was beginning to recognize as well.

The day after I got this note from gumby, a colleague came to observe my class. The class session was solid. I taught like me. I incorporated mindfulness. My colleague and I talked afterward. She was very complimentary. She compared the class and level of student engagement to a graduate seminar. I pushed back, asking her about my teaching persona. (I *always* ask about my teaching persona.) She laughed lightly and said it was great. I pushed on, telling her about my doubts and insecurities as a professor. She looked at me frankly and said, "You need to change your perspective."

In that moment, I could see the arc of my recent life. Between the age of twenty-one and thirty-one, I was experiencing things that everyone goes through: heartbreak, bad health, dejection. It had worn me down, my resources were depleted, and I needed space to recover, to devour life again, to grow fat and strong again.

For a long time, I had been repeating a mantra—that I wasn't the right fit, that I wasn't professorial—but I didn't even believe it anymore. I had been stuck in the valley, but all I had to do was stand up and walk to the crest of the hill to get a breath of fresh, freeing air. I replaced that self-pitying mantra with one that had served me well in happier times: "Trust thyself."

For a time, I was able to pour myself into my work, without reservations. Even my students sensed it. They commented on my passion; they liked the way I put myself out there. They were learning, not just about academic writing and research but about themselves.

Things went well for a while. But those old struggles would return in my next position, and I began to doubt whether I could really change things from within the system.

---

gumby and I would see each other occasionally over the next seven years. I eventually learned his real name, and I even put together his genealogy, more to assuage my own curiosity than his. gumby was descended from a long line of folks from the eastern United States. That didn't really matter, though. What stood out was that gumby's parents were bank robbers.

One cold February morning, gumby invited me for a walk. I needed it, badly. I was feeling beaten down by the frequent changes in leadership and the entrenched hierarchy in my department at the university.

The day was gray and so was my mood when I pulled off to the side of a gravel road to park my car. gumby and I had agreed to meet at a local nature preserve, a rich hardwood forest that spanned the bottomlands of a frequently flooded creek with steep slopes and rolling diabase hills.

When I pulled up, gumby and his dog, Sherlocke, were near his beat-up red van. The two rear doors were open, revealing a large open space for sleeping and some crates held down by cables for a few important items, like toiletries and books. gumby had just moved out of a trailer and into the van. He was going to live a vagabond life.

gumby and I greeted each other awkwardly with half smiles. It had been two years since we had last seen each other. We began to do what we often did, walk. gumby led the way into the rich bottomland forest. Evidence of recent flooding was everywhere: dense leaf packs were jammed into the base of trees thirty yards from the bank, new grayish-hued sand was deposited over the forest floor, and a giant port-a-John lay on its side, door open, washed up or down from who knows where. The port-a-John tickled gumby. It made me cringe.

We walked closer to the creek until we found a one-foot-diameter fallen tree, submerged in the middle, that more or less spanned the creek's width. gumby walked across barefooted. I was wearing my leather hiking boots, and while I was prepared to fall into the icy stream, I really preferred not to. At one point, while crossing a wet, slippery part of the log, I stuck my arms out to both sides to quickly catch my balance. At another point, I actually got down on my hands and knees to cross over. That made gumby laugh. The fallen tree didn't quite reach the bank on the other side. gumby took a long stride. Sherlocke had swum across. With my short legs, I took a leap and wondered how I'd get back.

On this side of the creek, the landscape was more dramatic. We scrambled up a slope so steep that gumby and I used trees as handholds to pull ourselves up. By the time we got up, we were breathing hard. gumby asked if we should sit and chat near a large oak, and we did. At first, we just sat. gumby was catching his breath, and Sherlocke was putting on a show, pulling branches across the ground, running back and forth, sometimes coming near to be petted and praised.

Eventually gumby and I began to talk. I told him that I was worn down and becoming jaded. I was wondering if gumby had a cure. He did. gumby said he had a five-step plan to help us, and society as a whole, emerge from our insane *wetiko,* a Native American word used to describe the greed-sickness of colonizing Europeans.

gumby's first step was to recognize that we have a problem. He said that that's where so many are stuck today—they're stuck thinking that our society is the greatest there ever was, and that doing more of what we do would only make it better. gumby thought that was bullshit. He saw our society as one that trapped us in cycles of need, greed, and obligation. I was inclined to agree.

The next step for treating *wetiko* was to change our economic system, first by becoming scavengers and then by re-creating a sharing, hunter-gatherer economy. gumby said that for those who first began this journey, they could easily live on the excess and refuse of American consumerism. That was what gumby was doing now. He didn't have a job, hadn't had a steady one for nearly a decade. He also didn't have a house, or any of the bills that went with it. He dumpster-dived for his food, went to the library for books, and traded his knowledge of survival skills for anything else that he needed. Every once in a while, he was paid for his teaching. gumby said that the only time this lifestyle was difficult was when it came to medical care. He had suffered from a few serious illnesses in recent years, pneumonia and a gall bladder infection, and he couldn't afford to go to the doctor regularly. gumby said that he knew, in the long run, that this would probably shorten his life, but it was worth it to experience relative freedom from *wetiko*.

The third step in gumby's plan to release us from *wetiko* was to change the government. gumby advocated for anarchy. My face must have registered confusion, because gumby felt compelled to explain that most people misunderstood anarchy. In gumby's conception, anarchy was the type of government embraced by tribal societies, a government where natural leaders, typically elders, guide the tribal members but never force or enforce their guidelines.

Next, gumby advocated for a resurgence of animism, the idea of seeing all things as "people." gumby said that he was still working on this himself, but he was practicing by acknowledging all life forms as people, for example he would talk about "tree-people" or "fish-people." Animism was gumby's way of ensuring that all organisms, and the entire natural world, were respected and recognized for their inherent value. Animism was a way of creating equity and sustainability.

At this point, we had been talking for forty-five minutes, but the last step in gumby's *wetiko* cure was unclear to me. Perhaps its central tenet was redeveloping our "primitive" or survival skills: learning how to build fires, tie knots, make traps, build shelter, forage, navigate, and find water.

By now, gumby and I had left our place by the tree and had moved west toward a man-made lake. Sherlocke jumped into the water again and again, chasing large sticks that gumby would throw out to him. We found otter scat, blue hued and full of fish scales. We found deer tracks and deep-set raccoon tracks. We clambered over more logs. gumby and I had talked for a long time, but not long enough, never long enough. I took my shoes and socks off to cross the creek; gumby continued barefooted as we walked along the gravel road back to our cars.

Eventually, we returned to the grassy parking area. We hugged goodbye, a long hug, the type that one gives when you don't know when or if you will see each other again. I left vibrating, elevated and energized, challenged and recharged, nervous and hesitant. It was time to make a change. I got in the car and drove north toward home, past hundreds of businesses—bookstores, car lots, pet stores, wine shops—none of which gumby ever frequented, unless it was to scavenge food from the dumpster. I came home and opened the windows wide, letting the brisk breeze cleanse the house.

I suspected that gumby was right about everything. The world is so skewed toward greed that we don't see how mechanical, cold, and cruel we have become. We drive to work each day. We see a person in need or a car that's broken down. We don't stop. We have to get to our job on time or risk losing it. We have homeless people in our own communities, we pass them every day, but we don't solve the underlying problems keeping them in homelessness.

As I reflect about gumby, I see that he has an appeal to me that is similar to the beautiful Scarlet Kingsnake, like the one tattooed on his hand. Both gumby and snakes, I suspect, are scary to some and misunderstood by most. And yet they are both independent. They both can leave chaos behind and slip away into the water, into the woods, and into the night.

THE NEXT GENERATION

IT IS HARD TO predict how snakes will be viewed by coming generations. Human populations are becoming increasingly urban, and today nearly 80 percent of Americans live in a city, half of them in just three major urban areas: New York City, Los Angeles, and Chicago.

These cities aren't like the Chicago that my father lived in the 1950s, a Chicago dotted with empty lots of remnant untilled prairie filled with dense populations of garter snakes, garter snakes that would end up in coffee cans or spread out in front of a three-year-old wild child who thought he might train them to perform in the circus. No, the urbanization of the future means that people won't interact with snakes very often or ever. But it's unclear what that lack of visceral, daily experience with snakes means.

Consider more rural populations, perhaps like a family I know living on a small farm in middle Tennessee with peach trees, hoed rows ready for tomatoes each year, a barn and pasture for horses, a pen for chickens, and a huge garage for the retirement RV, a farm surrounded by lush deciduous forest filled with the widest sassafras trees and the tallest oaks, bordered by a rock-bottomed creek and limestone bluffs dripping with ferns. In this idyllic setting, venomous snakes are hardly ever seen; instead, harmless Black Rat Snakes hunt for mice in the barn, and watersnakes sit at the edge of the cold creek catching silvery shiners. Yet, when they are seen, the woman of the house with wild raven hair yells, "kill it, kill it!" If she looks around and her husband isn't home, she will run, grab the hoe, and chop the harmless snake to bits. If her husband is home, he might "take care of it." Sometimes taking care of the snake means moving it someplace else, sometimes it means killing it, the outcome depending on the mood of the man and the temperament of the snake.

In these more rural American populations, proximity to snakes does not necessarily engender respect or understanding; often "familiarity breeds contempt." Unfortunately, vicinity alone doesn't create connection, but connection with snakes—a deep, abiding sense that one's past, present, and future are deeply intertwined with that of snakes, that our lives will mirror the snake's own life—is what is needed to protect them and us.

Connection to snakes requires a cultural shift. The same inner machinations that justify racism or rudeness, bullying or brutality, callousness or cruelty to our fellow human beings, the same justifications that we use when we refuse to make eye contact with the beggar in his orange vest carrying a cardboard sign on the corner or when we roll our eyes and shift our weight in frustration when we're stuck behind an ambling elderly woman with a cane in the tight cereal aisle at the grocery store or when we lay on the horn when the driver ahead of us hasn't stepped on the gas quite quickly enough after the light turned green, those same justifications also operate when we are unkind to animals. That same brutality, callousness, or cruelty is operating within us when we leave worms to desiccate on the sidewalk, or smash a spider on the bathroom wall, or shift the steering wheel just enough so that our front tire hits that snake sunning on the side of the road. This is the modus operandi of a culture fixated on time, money, and the cult of personality rather than the golden rule and communities of caring.

Our current Western culture doesn't particularly embrace the wisdom traditions, the most enlightened tenets developed by cultures around the world. We often ignore the underlying lessons of Hinduism, Buddhism, Islam, Confucianism, Christianity, Judaism, and we're even further away from understanding the wisdom of the animistic religions of Native Americans, Shinto practitioners, and pagans. Many of these traditions embrace behaviors that benefit others, believe that life is lived in relationship and that respected others aren't limited to people. They also uphold basic values like truth and kindness. These traditions strongly emphasize a life of connection.

If we shift our gaze to the world of education, we find that same lack of connection. Even in professional programs in the environment, programs offering master's-level education in managing the environment, this is still the case. Often students come into these programs with little knowledge of or connection to the natural world. No matter where we grow up—in urban, suburban, or rural environments—we all should have the opportunity to learn that wispy, sheet-like cirrostratus clouds precede rain and that a full moon gives off ten times more light than the first quarter moon (the phase that looks like a half moon), and we should all know the songs of the ten most common birds because birds are everywhere, even in the city. As Richard Louv and others have taught us, kids today can identify more than one thousand corporate logos but only a handful of local plants and animals. That deficit often doesn't change as people grow up.

In a School of the Environment, students aren't typically hugging trees or singing in the rain or talking with snake spirits. In fact, the same prejudices we find in the general population exist in the population of those dedicating their lives to environmental sustainability and justice. Among those students who do gravitate toward nature study, they tend to quantify mammal habitat corridors or differences in bird species richness or perhaps even salamander distributions, but very rarely does anyone study snakes. In the last twenty years, with hundreds upon hundreds of students coming through our Master of Environmental Management program, only two master's projects have explicitly and solely focused on snakes.

I once mentored a master's student with broad interests, Diego. Diego came to the study of the environment by way of an economics degree. He cared a lot about people, he was active in student council and the Black and Latino Club, and he wanted to advance diversity, equity, and inclusion in our school and beyond.

Diego had an open heart and an open mind, and at some point during his time at Duke, he became fascinated by snakes. At first, he thought he'd try his hand at fieldwork. He considered monitoring transects—lines of tin and plywood coverboards—set out in the seven-thousand-plus-acre Duke Forest for snakes, but the terrain was tough, the vegetation thick, and data hard to come by. So he shifted his attention to his fellow students and began to investigate two questions: (1) What were the current attitudes toward snakes within the professional environmental master's degree community; and (2) What new methods might decrease anxiety and fear toward snakes within that community?

Diego's work was premised on that fact that snakes provide valuable ecosystem services and that for this reason they should be protected, but that protection was less likely to happen if future policy-makers (like these master's students) couldn't see the objective value or benefit of snakes.

Diego found that people's career interests were a strong predictor of attitude toward snakes. Now remember, all these students were studying the environment and chose to follow career paths that protect environmental resources. Still, those students more interested in energy, economics, and policy held more negative attitudes toward snakes than those whose degree focused on environmental science and conservation. Also, there was a clear correlation between people's chosen career track and their frequency of outdoor recreation. Those students who reported being outside more also had more comfort around snakes.

Diego's results didn't vary that much from what we know about a related concept: positive environmental identity development. Environmental identity is a tricky idea, one developed in response to the complicated and indirect roles that environmental attitudes play in shaping pro-environmental behaviors. The research indicates that having pro-environmental attitudes—that is, having concern for the environment—doesn't necessarily mean that one will act in ways that benefit the environment. But if one has a strong environmental identity, one is more likely to actually engage in pro-environmental behaviors, like recycling, taking the bus, or buying a low-flow showerhead.

Environmental identity is in part influenced by what researchers call "significant life experiences." For those who have strong environmental identities, these significant life experiences usually occur in childhood or young adulthood and include time spent in nature, some sort of study of natural systems, and witnessing environmental destruction, habitat loss, or injustice.

We also know that environmental identity is influenced, in part, by our social environment. Children are influenced by the adults around them, and if a parent or teacher or an important mentor has a strong environmental identity, then it's more likely that the children they interact with will too. While family is a particularly strong influence, friends and even coworkers can help us develop our environmental identities at least into early adulthood.

Pro-environmental behaviors are also related to empathy. If someone has developed empathy for animals, for instance, they are more likely to donate to causes that support wildlife or switch to a plastic-free lifestyle to save sea turtles or put out a bowl of water for animals during drought conditions. To develop our capacities for empathy and compassion, we need connection, or, as researchers call it, attachment. Attachment in early childhood—those bonds of affection and love we experience—affect us for the rest of our life. Secure attachment between a child and caregiver can offer protective psychological benefits for the child's entire life, influencing mental health, well-being, delinquency rates, empathy, and pro-social behaviors.

Even the bond between child and pet has profound consequences for human development, offering stability, reassurance, and protection. Bonding with pets can also help kids who didn't have secure bonds with caregivers learn to develop those bonds. Moreover, positive attachment to pets predicts positive attitudes toward other animals, and some research indicates that this can transfer to snakes. Once our capacity for empathy is triggered, we can cultivate it and allow our sphere of empathy to grow, but if we don't have secure attachments, our capacity for empathy becomes stunted.

With snakes, Diego was finding a similar chicken-and-egg scenario. Those who had connection to nature also exhibited more pro-snake attitudes—they were less fearful and more comfortable interacting with snakes. Oddly, those who had pets weren't necessarily more comfortable with snakes. So the question became, how do you create conditions for those who are unconnected to and fearful of snakes to actually interact with snakes in a positive way? Diego tried to answer this question too. In his survey of students, Diego proposed six interventions to increase pro-snake attitudes. Some of these interventions were passive, such as seminars about snake biology or cultural views of snakes. Some of these interventions were active and involved learning how to handle snakes safely. The surprising result was that the majority of participants preferred interventions that involved handling the snake in some way, although a large number also preferred just to learn about their benefits.

For a group of people who weren't particularly comfortable with snakes as a whole, they were sure interested in learning more about them. Diego's work was the talk of the town, with students popping into my office asking about his research and if he was really doing a master's project on attitudes toward snakes. There was a tremendous buzz before his public presentation, and the day Diego gave it, his presentation was so heavily attended that people were standing at the back of the room. Yet there was also a kind of jocularity in the air, almost like studying snakes and the attitudes people have about snakes was so odd that the presentation itself would be some sort of spectacle.

But the project was built on solid methods—intense survey work, intense statistics—and Diego presented the work in that way, as serious, important. When Diego finished his presentation, the applause was thunderous and the questioning intense. People were fascinated, they wanted to know more, and they wanted to know what it all meant.

There is something to be learned from this, from the spectacle and fascination. It showcases the profound disconnection between people—even students studying the environment—and the natural world.

When I teach, I frequently take my students to a planted Longleaf Pine stand, about fifty miles outside of the longleaf's native range. Last October, I told my students about rare Red-cockaded Woodpeckers forcing holes into rot-softened trees, drilling wells to create a sometimes smooth and sometimes sticky shield around their aerial apartment. I talked about Black Rat Snakes climbing fifty feet up a straight pine, hoping to make a meal of woodpecker eggs or chicks, only to be stymied by the shield, falling to the ground.

While snakes are excellent climbers, they do fall. They fall from Longleaf Pines, and they once fell from ancient Roman roofs, which was seen as a particularly bad omen. Imagine a snake falling through your *conpluvium* (an architectural hole in the roof of your house) and into your *impluvium* (rain collection basin), and the next thing you know, a wedding needs to be canceled. From Mesopotamia to ancient

Rome, snake omens foretold social discord. In the snake world, a falling snake often suggests a more complicated discordance, a mismatch of form and function, of ambient conditions and biological traits.

One reason snakes fall is from the cold. Corn Snakes—semi-arboreal, heavy-bodied snakes found throughout the American South—are at least ten times more likely to fall when the weather gets cool. Since snakes are ectotherms, their ability to move is affected by changes in temperature. When it's cold outside, snakes move more slowly and change their shape, looping more around branches rather than stretching out. The looping helps snakes balance on thin branches but isn't effective on thicker limbs, making them more likely to lose their balance on thicker surfaces and fall to the ground.

Truly arboreal snakes, like the Green Vine Snake (*Oxybelis fulgidus*) and Blunthead Tree Snake (*Imantodes cenchoa*) of Central and South America, have different body plans than other snakes, with the bodies of arboreal snakes being lighter and longer. This means that the weight of the snake is stretched out across a branch rather than focused on a particular point. Long, thin-bodied snakes can also make more loops around tree limbs, improving balance. Plus, this body plan allows arboreal snakes to stretch across the gaps between boughs. Arboreal snakes also have other adaptations for life in the trees. Their scales are different, their bones are different, and even their heads are different.

Snakes also fall out of trees because of predators. While scientific studies on this topic are lacking, many herpetologists can attest to watching snakes drop from branches when birds or people come too close. Sometimes, the snake doesn't drop soon enough. Herpetologist Michael Plummer relates observations of Blue Jays carrying Rough Green Snakes to a tree, "peck[ing] them to death, and eviscerat[ing] them as they ate." These snakes live in the southeastern United States and probably spend more time off the ground than any other snake in the region. They hunt in shrubs and vines; they even court and mate off the ground. If everything is going well, the male will spy the female from afar. He comes at her quickly, head jutting back and forth rapidly and tail moving from side to side too. Then the male flicks his crimson tongue over the female's back, their bodies align as the male rubs his potential mate's back with his pale-yellow chin. After this, the male's tail moves, trying to align with the female's vent. Sometimes, a female green snake isn't too keen on the male's advances and might jerk her entire body when the male touches her and slide away a short distance along the branch.

Myths abound about falling snakes, especially in the American Deep South. Legend has it that boaters are at particular risk from attacks by Cottonmouths falling into the boat. Cottonmouths are heavy-bodied snakes that rarely climb, but their harmless, distant cousins—the watersnakes—often bask on limbs. And watersnakes do fall, especially, as herpetologist David Steen suggests, when needing an escape plan after being startled by a passing boat.

This scenario can be avoided. If we're aware of our surroundings and respectful of the creatures with which we share habitat, we are less likely to have startling encounters. If we show young people how to appreciate the natural world and approach it with a sense of wonder, they will be more likely to react with empathy to natural phenomena. If we focus on connection and embrace the variety of life-forms around us, a falling snake can become a point of meaningful interspecific contact. For me, seeing a falling snake isn't a bad omen or a source of fear, but a sign to be more present in nature and press on.

And that's what the next generation is doing, pressing on.

As a budding snake conservationist, one can explore the realm of social attitudes toward snakes or focus on snake biology and ecology. Take the Eastern Diamondback Rattlesnake, for instance, a hefty snake with a mosaic tilework of tan inside black inside cream diamonds along its back. This is a snake that exhibits Zen-like stillness hunkered against a fallen log in the woods, until that tail starts rattling like a wagging finger. The pervasive social attitudes that label Eastern Diamondback Rattlesnakes as dangerous or evil have devastated their populations by perpetuating practices like rattlesnake roundups. But the Eastern Diamondback Rattlesnake also experiences pressure from human behaviors not aimed purposefully at its destruction, such as the threats of habitat loss and degradation. This is the realm of snake biology and ecology.

For many snake species, the importance of habitat can't be emphasized enough. The Eastern Diamondback Rattlesnake in particular—the largest rattlesnake species in the world—used to be found along the Coastal Plain of the entire southeastern United States, from North Carolina to Louisiana. Today, one of the primary habitats of the Eastern Diamondback Rattlesnake has nearly disappeared—the Longleaf Pine forest. Less than 1 percent of the original ninety-two million acres of Longleaf Pine in the United States remains.

Longleaf Pine primarily disappeared because of the naval stores industry, an industry that extracted tree products like wood, resin, pitch, and turpentine, to build and maintain large wooden ships. During the early 1600s, Britain, having already lost its forestlands, relied on Sweden for its supply of naval stores. But in the mid-1600s, the price of Swedish products doubled, and England turned to the New World to meet its needs, needs met by the destruction of White Pine forests in the northeastern United States and Longleaf Pine in the mid-Atlantic and southern states.

With so much of its former habitat destroyed, conservationists are working frantically to protect the Eastern Diamondback Rattlesnake, mostly by protecting its habitat. This includes the Jekyll Island Conservation Department in Georgia. Here, a team of folks has been using radiotelemetry to track Eastern Diamondback Rattlesnakes for nearly a decade. The result is a rich dataset on the movements and habitat use of a rare species.

Jekyll Island is a narrow, oblong bit of Georgia, stuck between Savannah and Jacksonville, Florida. The 5,847-acre barrier island—one of fourteen that line Georgia's coast—has a strange, machete-like shape, with its broad end pointing north, jutting into Saint Simons Sound, and its handle pointing south. Development of the island is concentrated in the center, and recently researchers found that human activity had essentially split the Eastern Diamondback Rattlesnakes on the island into two populations: a north population and a south population.

My graduate students Kelly Joyner and Hannah Royal decided to work with the Jekyll Island Conservation Department to see if those populations could be reconnected. The aim was to use the radiotelemetry data in conjunction with a plan to close down at least part of a Jekyll Island golf course to create a viable habitat corridor connecting the populations at both ends of the island, allowing those isolated populations to interbreed and increasing genetic variation to ensure the long-term survival of the populations.

Today, 40 percent of Jekyll Island is covered fairly evenly by just four land cover types: maritime Slash Pine and Longleaf Pine upland flatwoods, dense development, golf courses, and roads, and those proportions don't bode well for snake survival. The dense development means more potential for human-snake encounters, which often end with a thwack and a needless snake death. Golf courses limit snake movements, and some species, like Black Rat Snakes, accidently swallow golf balls, mistaking them for bird eggs. Plus, snakes are excruciatingly vulnerable to cars as they thermoregulate on or simply cross the road.

But it's not all bad news. Hannah and Kelly found that the northern forest-dwelling population and the southern marsh-dwelling population of Eastern Diamondbacks could see improvements in connectivity if even just one of the four golf courses was returned to maritime grassland or longleaf savanna. This is an encouraging finding because that was the plan: one of Jekyll Island's four golf courses (currently covering 10 percent of the island) needed to be retired.

It turns out that golf courses are closing around the country, with nearly one in ten courses in the United States having closed since 2006 alone. Kelly and Hannah's search in the literature and popular media showed that between aging player demographics, the expense of the sport, and the rise in popularity of other sports, golf just couldn't compete. According to Kelly and Hannah, Jekyll Island is "no different."

So what are we to do with all these closing courses? It's easy for developers to swoop in and build houses, yet these closed courses also represent an incredible opportunity to reclaim habitat for wildlife and make greenspace accessible to communities.

Kelly and Hannah were able to outline a plan to do just that, with the added bonus of protecting the threatened Eastern Diamondback Rattlesnake. The plan, though, is complicated. The soils on golf course, especially ones as old as those on Jekyll Island, built between 1922 and 1975, are degraded. They've been compacted by years of treading feet and heavy maintenance equipment. This soil is tough on native plant roots, so the first step is to remove the turf grass and till up the ground. And unless native plants can be quickly planted and managed, restoration ecologists will also need to manage the long list of invasive plant species that will quickly march onto the freshly tilled earth.

Once the old golf course site is ready to be planted with natives, careful attention must be given to the plant species selected because it turns out that Eastern Diamondback Rattlesnakes have some preferences of their own: they prefer a fair amount of shrub and vine cover but less canopy cover. Their habitat preferences look a lot like the habitat structure of the Longleaf Pine ecosystem. While much of the golf course can be planted with maritime grasses, providing cover, over the long term, the site would need to be managed to balance the aesthetic needs of Jekyll Island residents and the habitat needs of the rattlesnake.

The need to balance the preferences of both residents and wildlife means that conservationists must focus on critical habitat areas. Kelly and Hannah recommended that the Jekyll Island Conservation Department direct attention to "pinch points," or narrow, but important, parts of the corridor that can connect the two Eastern Diamondback

Rattlesnake populations. These pinch points often occurred along roads, suggesting that wildlife underpasses could be a prime way to relieve the bottleneck.

While the work on Jekyll Island is still in progress, hope can be found. As we learn more about the importance of childhood connections to the natural world and to animals, and as more and more research becomes available on attitudes toward snakes and how to conserve these often too rare species, we still have the opportunity to protect remaining snake populations. Kelly and Hannah noted that it "may take decades of dispersal" to reconnect the two populations of Eastern Diamondback Rattlesnake on Jekyll Island, but the possibility exists and dedicated folks in the Jekyll Island Conservation Department, with community support, can make it happen.

WORDS AND WISDOM

 I HAVE LIVED IN the Piedmont of North Carolina for nearly two decades. Here, summer reaches across the year like a hot, sleepy dog, arching and stretching, paws digging into the sod. In the Piedmont, spring ends prematurely, and autumn waits nearly until winter before relieving the aestival sultriness.

By September, the leaves of southern oaks are sere and tattered. Breaks in the canopy of Loblolly Pines and mixed hardwoods reveal the white-pappused heads of Fireweed or great stands of Yankeeweed, that fennel-sweet symbol of a civil war gone wrong for agrarian enslavers, and gone right for the cause of freedom, as it filled fallow fields in the southern Piedmont after the Yankees marched away.

By September, the heat is bearable, the humidity of midsummer tamped down, and the average highs in the 80s (°F). This makes for good snake weather: in midmorning and early afternoon, when the thermometer reads in the low 70s (°F), snakes are often seen in a patch of sun that snuck past the canopy of the secondary-growth forest.

Plus, the snakes are moving this time of year. In southeastern North Carolina, even snake species whose populations are in decline, like the Southern Hognose Snake (*Heterodon simus*), are seen most commonly in September and October. It's hard to know what snakes are up to in September, but radiotracking and careful observation can tell us a lot. Jeff Beane at the North Carolina Museum of Natural Sciences and his colleagues observed a pair of Southern Hognose Snakes mating in September, and dead-on-road snakes with lots of lizards in their stomachs suggest that they were also focused on eating.

Other studies show the seasonality of snake movements. Herpetologists Whit Gibbons and Raymond Semlitsch summarize them well, explaining how some snakes in the temperate United States have two peaks in activity during the year, during spring and autumn. This is also known as a bimodal peak of activity. Some snakes just have one peak in activity (unimodal); this usually happens in midsummer, but as is the case with the Southern Hognose Snake, it can happen later in the season too.

Snakes move for all sorts of reasons: to disperse from hatching sites, find sanctuary from the heat or warm up a bit, avoid predators, seek a mate, and locate a spot to lay eggs. During the summer, snakes might be seen midday when it's cooler or searching for food during those crepuscular hours to avoid the midday heat on hotter days. The exact timing of activities depends on the species. Midmorning, Rat Snakes might be in active search mode, crawling along the forest floor, poking their nose into the nooks and crannies of fallen logs or climbing up a tree searching a nest for eggs.

I recall walking along the bottomlands of the Eno River in mid-July one morning with my family. We heard a kind of terror in the call of a bird. When we looked up, there was an Acadian Flycatcher lunging aggressively at a Black Rat Snake, but never coming closer than a couple feet. The Black Rat Snake, at least ten feet up in a young Swamp Chestnut Oak, was consuming the small clutch of not-quite-one-inch-long eggs, creamy ovoid jewels with brown beauty marks.

In temperate climates, snakes can move long distances—typically up to a mile—to and from hibernacula, or dens, where they overwinter. Often, snakes will come back to the same den year after year, overwintering with snakes of the same, and sometimes different, species. One of the most famous overwintering sites for snakes is in Manitoba, Canada, where Red-sided Garter Snakes fill up large, underground caverns formed from limestone bedrock. Herpetologists will trek to the site in Narcisse in late April—which also marks the start of mating season—and then again in early September to see the snakes leaving and entering their winter den.

In Durham, North Carolina, where Duke University is located, early fall marks another migration, that of students to campus. Soon small groups of undergraduates are asking to be taken on an exploration of the Piedmont woods, searching for snakes. Not lizards. Not toads. The goal is snakes, the allure of an animal deemed slightly scary and decidedly unfamiliar.

Typically, I meet the students at the edge of the woods, our cars parked along a hidden gravel road that dissects one of the Piedmont's premier preserves or valuable working forests. We talk about safety. The safety of the eager students. The safety of the surprised snake.

Training by an expert is important, since people often use too much force when handling a snake, failing to adapt to the snake's needs. Even my preferred technique can go awry and break a snake's neck if someone isn't trained well. For novices in particular, hoping to handle some of the large-bodied snakes we find in the Piedmont, like the semi-arboreal Black Rat Snakes or sleek Black Racers, I recommend an approach of gentle control: secure the head by first pressing a snake hook and then your fingers, very gently behind the neck—back off if the snake thrashes, and as you lift the snake's surprisingly fragile body, support it around your arm.

The technique works. Students don't get bitten, and they develop a first-person, intimate respect for snakes. And, if the students are coached and trained, snakes don't get hurt, released without lost teeth or a broken back.

One September, George—who had an encyclopedic knowledge of snakes—accompanied our group into the field. I began to give the students my own tried-and-true advice about snake handling, but George interrupted chidingly, saying, "Now here, I respectfully disagree with Dr. Cagle," although the interruption and subsequent wholesale takeover of the excursion hardly felt respectful. "In my experience, herpetologists generally hold snakes by the tail first. The snake is calmer that way and the head is away from you." I cringed internally, but I didn't argue with his right to respectfully disagree.

That day, we searched a typically productive patch of xeric, mixed hardwoods and came up empty. The students returned to their vehicles after forty-five minutes, and we began to caravan along the gravel road, heading to another site with George in the lead car. Soon the cars stopped. George jumped out and grabbed a beautiful, five-foot-long Black Rat Snake. The students jumped out of their cars too, immediately snapping photos on their cellphones, and George began to talk about snakes.

George was senior to me. His teaching career about doubled my own, and he had a professorial way of holding forth that I never could successfully ape. My own approach to teaching in the field was more sensorial, quieter. I explained how to identify snakes when we found them, and I shared a couple interesting tidbits about their ecology, but I preferred to let students observe, respect, and reflect.

I suspect George would respectfully disagree with that approach too. Instead, George talked and explained and expounded. He touched upon snake anatomy: *Did you know that the eyelid of a snake is really an ocular scale called the brille and it sheds with the rest of a snake's skin?* And snake evolution: *Snakes likely evolved from burrowing lizards and one of the oldest snake fossils comes from Colorado dated to about 155 million years ago, the Upper Jurassic, when "the house cat croc," a three-foot-long crocodyliform roamed the area.* And snake behavior: *Snakes*

yawn not because they are tired but rather to gather a huge pocket of
air to pick up chemical clues about the environment and analyze it with
their vomeronasal organ, also known as the Jacobson organ. He touched
upon quite a bit. His words were like a flood. There were just so many
of them.

I once read a book where a Native American elder said, "People
should think of their words like seeds. They should plant them, then
let them grow in silence. Our old people taught us that the earth is
always speaking to us, but we have to be silent to hear her." The elder
went on to describe *wasichu*'s (white people's) use of words, saying
that to Indigenous people it "just sounds like a bunch of people saying
anything that comes into their heads and then trying to make what
they say come around to something that makes sense."

I felt that way now with George. I felt that his words had turned the
snake into a circus freak, a biological and evolutionary alien for us to
poke and prod and provoke.

George's words continued. He twisted the snake upside down to
show its vent, to guess its sex. He chastised me for not bringing snake
probes, to poke and prod the snake for real, so his curiosity about its
sex could be patently assuaged. He passed the snake around, into six
to eight different pairs of hands. The circus performance continued,
and George had planned the grand finale: press the big black snake
against the bark of a tree to show how its body conforms and climbs,
while telling the students about snakes' aerial predators—passerine
birds and hawks—as we walked back to our cars, leaving the snake
vulnerable to those same passerine birds and hawks.

I was numb with cold fury. I wanted to yell, to put a stop to behavior that showcased deep knowledge but shallow respect. We had
objectified that proud, elegant snake. We had terrified it and stressed
it and possibly endangered its life. This was not the way to teach conservation. This was the mindless way that has led to severe declines in
snake populations in the first place: rattlesnake roundups, captures for
the pet trade, land development, highways dissecting every potential

parcel of habitat. We know so much, but non–Native Americans have long had difficulty applying that knowledge in a way that respects life, beauty, or wholeness.

I stayed silent. That silence was part wisdom and part cowardice. I was too angry to share my thoughts calmly or clearly. I was also at a disadvantage. I could never call forth the flood of words with which George would surely meet mine. The words I could use to explain my position would be simple and few.

We drove to another site in the Piedmont woods, a mesic hillslope with a denser canopy of hardwood trees, like the beech with its smooth, lichen-mottled bark and long, outstretched limbs that cool the forest floor.

George led the students down the slope. We flipped cover objects— boards and logs—in search of our prey. I stood by a large piece of tin with three young women. We lifted it up and underneath lay a slender, smooth Black Racer with the slightly milky look of one about to shed its skin.

I used my method, pressing the hook gently behind the snake's head, reaching down with my hands and replacing the hook with my fingers, keeping the body calm and still as I draped it over my arm. We called out to the rest of the group. I began to introduce the snake to the students in my few-word way. Look. Observe. This is a Racer. See its minute scales, smooth. Not like the Black Rat Snake. See the milkiness. It's about to shed. It can't see as well now. Awe. Wonder. Silence.

George came up next to me. The silence needed to be filled. He took the snake from me. He said something about being afraid and about real herpetologists not needing to secure the head. My ego bruised and bristled. I had been bitten by more than one hundred snakes in my career; I had devoted huge swathes of my life to the dogged study and conservation of snakes. I mumbled something weakly.

George's flood of words began afresh. The students gathered tightly, taking photos, reaching to touch the snake as George talked. A young woman from China, who hadn't seen wild snakes before, leaned in closer. In a flash, the snake struck, biting her hand. She pulled back. Her hand bled from a little red gash, like the scratch of a blackberry thorn.

George looked toward me, momentarily abashed. Our eyes met. We both knew that wouldn't have happened if he had secured the head. Instead, he looked at the girl and said that her bite wouldn't have bled as much if she hadn't jerked her hand back in fear. My jaw clenched at how deftly he had shifted responsibility onto the young woman.

I walked her back to my car, where I had a little first aid kit tucked away in the trunk. I quietly applied antibiotic ointment to her tiny bite, just to feel like I had done some good. I stayed at the car as she walked back to join the group.

The young woman from China had arrived that day as a potential advocate for conservation, open to developing deep respect for one of earth's most fascinating creatures. She left, instead, victimized by bravado and assaulted by words. We had lost our opportunity. I vowed never to let that happen again, and I drove away from the circus.

THE MIRROR

IT WAS LIGHT IN my hand. Lighter than a clump of Old Man's Beard lichen. Lighter than the dried leaf of an American Sycamore. As light as a petal, but less dense, a labyrinth of air.

It was frail. If I closed my hand, it would have crackled, cracked, and crumbled. Parts were sharp and brittle, like a dried-out sand dollar from North Carolina's Outer Banks. Other parts were soft and smooth, delicate like silk. I traced my index finger along one edge; it was curved like the bowl rim of a well-polished spoon. I placed the tip of my little finger into a rounded divot as tiny as a tea-cup in a child's meticulously furnished dollhouse. Long sections had a rough, regular, but intricate pattern, like running one's fingers across an ancient tapestry.

The color varied. In places it was bone gray. Other sections were a lighter elephant-tusk ivory. At one point the colors were surreal pale copper and an elegant champagne. I looked again and realized that in places it was completely transparent, and the color was that of my own hand seen through the lens of a shed snake skin.

I was drawn to the snake skin on the table that day. Eight women sat on chairs in a circle, gazing in toward an altar of nature laid out on two rickety card tables covered by a watery, turquoise batik. The altar displayed an apricot-hued shell with regular ridges, a scute-less bleached box turtle shell, a smooth, contorted piece of driftwood, and a dozen other well-loved objects that a wise Harvard-educated materials artist had collected during her long, roaming life.

We each chose an object that spoke to us, and we beheld those objects in silence. I was compelled, at first, to name, to identify the scientific family of the skin's antecedent, to pinpoint a specific epithet to quell the stark need of my science-trained mind. Were the scales keeled like a boat's ridged bottom or smooth like a tumbled stone? (Smooth.) Were the tail scutes "go[ing] through life two by two" or marching single-filed? (Two by two.)

The shed skin would have been a couple feet long had it been stretched out and entire. The spectacle—the shed of the clear lens that covers snakes' eyes—was hauntingly large, and the skin was tissue thin. It must have come from a Black Racer. I could see the racer in my mind's eye, black and sleek, alert as it raced toward the shelter of a large, hollow tree that had fallen several years ago.

And then I returned to the delicate skin. The scientist sated; the dramatist appeased. I was present again. With the surge of the conscious sea, I was pulled back into the depths by the undertow of the liminal zone. Inward again, deeper still, but not to the place of profound stillness. Instead, I was the skin. And I felt the dirty, dried-up layers of heart and soul falling away, my own ecdysis, allowing for renewed rejoicing in the natural splendor and richness of this sometimes love-filled world.

Strangely, I'm not alone in my fascination with the used skin of snakes. People use them in all sort of weird and wondrous ways. They make jewelry, the shed skin embedded in resin to craft a haunting cabochon pendant. Some people use it when painting their nails, lacquering the shed in gold polish to—I kid you not—"create a venomously hot look." Shed snake skin has been used in Chinese medicine, applied topically to heal sores and boils and taken orally for gallbladder problems and high blood pressure. Shed snake skins have also been used to model membranes in medical research.

People aren't the only recipients of the gift of a shed snake skin. Ground squirrels in California use rattlesnakes' shed skins to camouflage their own scent from predators. Some bird species also use shed skins in their nests. This protects eggs from predators like flying squirrels.

If the shed skin of a snake can provide so much value, it's not hard to imagine that snakes themselves are a crucial part of many ecosystems. Snakes are middle-order predators. They eat lizards, frogs, birds, mice, rats, even other snakes, and they provide food for hawks and owls and mesopredators like raccoons. Snakes also serve as pest control; without them, we'd have too many mice nibbling our crops. And don't forget, snakes reduce other pest species, like ticks. Plus, snakes are indicators of the health of other species.

In April 1935, Aldo Leopold—the renowned ecologist and progenitor of conservation science's "land ethic"—gave a lecture at the University of Wisconsin. It was the first place where he had ever used the term. The essay lays out, point by point, a clear and rational argument for a systematic overhaul not only of U.S. land management but perhaps of our society.

Leopold tells us that before the "machine age," human destruction of the landscape was at such a small scale with such relatively small impacts that the land tended to heal itself with time. He noted that this wasn't always the case and that we had seen societies collapse in places like the eastern Mediterranean, the Chinese interior, and, as later authors would add, in the New World, including the Mayan city of Tikal and the pinyon juniper woodlands of Chaco Canyon. In these instances, societies had taxed the land beyond its limit, past the point of sustainability. Leopold also asserted that our "present legal and economic structure" had developed before the machine age and in a place (Europe) where the land was relatively resilient to degradation by humans.

Most of us have understood Leopold's writing to mean that we need a new ethic toward the land, one which offers a set of moral principles that considers the land as an ecological or aesthetic whole. This is true and a fair interpretation of Leopold's work. But what is often left ignored is that Leopold wasn't just suggesting a new code of morality. In 1935, Leopold was suggesting something much more radical: an overhaul of our economic and political system.

Leopold saw that this system "contains the seeds of its own eventual breakdown." There was no mechanism to protect private lands from large-scale degradation that has tremendous public cost for present and future generations. Moreover, Leopold pointed out that there were no checks in place to identify and restore lands while they are only slightly degraded, when the costs of restoration were still low. Instead, the public inherited private lands that were so degraded that no one could manage to make a living out of the soil anymore.

The Carolina Piedmont can serve as a good example of this, particularly the old plantation (that is, forced labor camp) of ex–South Carolina governor William Henry Gist. Between 1832 and the end of the U.S. Civil War, Gist—a "fire-eating" secessionist—enslaved up to 178 people to eke product from South Carolina soil. In 1850 alone, Gist's Rose Hill plantation yielded 87,200 pounds of cotton, 7,000 bushels of corn, 2,000 bushels of oats, 600 bushels of wheat, plus barley, potatoes, butter, and livestock.

By the time formerly enslaved people took over the property after the end of the U.S. Civil War, the land was giving out. By 1936, the land was gutted. Erosion gullies deeper than a man is tall were scratched into the landscape. The land was too ransacked to be productive. So, the U.S. Forest Service—the public—took on the burden. The Civilian Conservation Corp (CCC) engaged in a "massive seedling-planting operation," and over the decades a forest grew back. But modern science reveals the scars on the land, the lack of organic matter in the soil. The land might support pine, but it is not a robustly functioning ecosystem.

Leopold was right. Our current system still fails to protect the land, and it fails to protect us, since we depend on the land for our survival. And just like our system has failed to protect topsoil from rushing to the sea or nutrients from being leached from the soil, our system has also failed to protect biodiversity. In the United States we have few mechanisms in place to incentivize private landowners to maintain diverse and thriving plant or animal life on their property. We have no mechanism to even fully quantify the diversity—the abundance and variety of species—on private land.

Without those economic and political incentives, social mechanisms, and a deeper underlying ethic that life is beautiful and should be encouraged to thrive, we also have no incentives to protect species. We have no incentives to shift our use of pesticides or grow native plants in our gardens for butterflies. We have no incentives to use UV reflective glass or turn the lights off at night for birds. We have no incentives to build a hibernaculum or retain a bit of cover or maintain a migration corridor for snakes. And it shows.

In places in Illinois where people used to "scoop them up by the handful" and "shake them out of the bed clothes," I saw no snakes. I've spent weeks or months or entire summers in places and not seen a snake. I visited South Korea—walking near the DMZ noted for its rewilding, climbing steep mountains leading to sacred sites, and visiting wetland habitats hopping with frogs north of Busan—and never saw a snake. I remember the overwhelming sense of discovery and joy I felt when I found one snake each in Japan and South Africa. This isn't the way these places should be.

Humans have eradicated one of nature's children from so many once wild and now tame places. It's a genocide as senseless as killing all the birds and smothering their song with the noise of highways and airplanes. It's a display of gross cruelty, abject greed, and supreme incompetence in shepherding the earth or even managing it well enough to enhance our true well-being.

For naysayers, it is true that no snakes means that we live without the uncomfortable humility of seeing a small slithering creature, one who knows the earth with an intimacy we can't imagine. We live without the humility of encountering a being with more power and potency contained in its small sharp teeth than even the richest among us obtains in an entire lifetime. But without that humility, we are ultimately left bereft.

We are bereft of reminders of our interconnectedness with all life. We are bereft of reminders of our mortality. We are bereft of forest friends and stories for our children. Without snakes—as friends or foes—we lose our humanity.

DISAFFECTED

 I HAVE NEVER ENJOYED the four- to six-hour drive between Durham, North Carolina, and Washington, DC. The land near the North Carolina–Virginia border is infused with a sense of abuse. The clayey should-be-subsoil sits naked near the surface. Years of intensive agriculture have washed away the once nutrient-rich topsoil, and the incipient, redeveloping organic layer is razor thin. Whatever old-growth forest was there is long gone, replaced by too many and too dense Loblolly Pines or young mixed hardwoods. The long stretches where one can drive under tree canopy should have calmed me, but I was always tense, wondering if I was about to run over the region's last box turtle as it emerged through the woods or hit a starving deer desperately crossing the highway in search of food. It's a bit of an overreaction— Eastern Box Turtles aren't going extinct. Yet.

I always had a sense that everyone on the road was driving too fast through a region whose history needed to be taken in slowly and deeply. This was a history to be wrestled with, to take into our souls, to gasp at and wonder at and ask ourselves why and how and then think about what it means for us today.

In this region sits a historic site, Fort Christanna. The site was devoid of people when I visited—just a field of invasive grass with monuments and interpretive signs—a site whose history was slowly decaying and whose past visitors had long forgotten its existence. Fort Christanna was the westernmost outpost of the British Empire in 1714, built to protect the colonizers and local tribes from the Tuscaroras. While the pentagon-shaped fort was manned and housed at least five large cannons, the "protection" mostly came in the form of "reeducating" the Saponi and Tutelo, Siouan peoples who lived nearby on allotted land. Eventually Fort Christanna came to be seen as an unnecessary expense that really served to protect the interests of the Virginia Indian Company. In May 1718, the colonial government had signed a treaty with the Tuscaroras. By 1740, most of the Saponi and Tutelo themselves had moved away, eventually adopted by the Cayuga.

In Virginia, the most-visited historic sites are in the middle and north of the state, and they are firmly rooted in post-European settlement history, emphasizing the history of colonizers: Historic Jamestowne, Thomas Jefferson's Monticello, and Appomattox Court House. I, myself, was headed to George Washington's Mount Vernon. I wanted to see the soaring old trees he had planted before they all died and his reconstructed distillery.

The tension of the drive, constantly feeling like I was dishonoring the places I sped past, was nearly too much for me. When I arrived at Mount Vernon, I blew past the entrance to the parking lot and had to make a U-turn. I finally parked and took a deep breath, barely holding back a cry of frustration. I had arrived.

Before my tour of the house, I stopped in the garden. It was teeming with butterflies—a new brood of bright and clean Eastern Tiger Swallowtails, mostly yellow, but some in that wondrous black morph. I walked the path around Washington's Bowling Green, dating back to

1787. This then-innovative landscape design had been made level with heavy rollers and was meticulously maintained by enslaved workers wielding scythes. I winced at the iniquity of early America and sought out something to recalibrate my soul: big old trees.

One of the most impressive trees at Mount Vernon is a Tulip Tree that sits along the Bowling Green and was probably planted around the same time the Green was established. The Tulip Tree had a wide diameter and stood impressively tall. Its gray bark lacked the regularized, anastomosing ridges that identify most Tulip Trees. In its dotage, the bark had become more deeply furrowed and mottled with the alternating knobby scars of old branches and whitish bare patches. Craning my neck until it hurt, though, I could finally see the canopy, still lush and green, with high branches big enough to be mature trees of their own.

Washington was known for his love of trees, having planted Eastern Red Buds, American Elms, and American Hollies around Mount Vernon. Did the enslaved staff of Mount Vernon ever fry up the redbud blossoms and eat the small, immature pods like garden peas? Some people today still do. Did Washington hold back tears when he transplanted a bunch of American Holly from the local woods and they all died? He cared enough about those young trees to get seeds from his brother. When that big Tulip Tree was planted, did he know that it served as a larval host for swallowtail butterflies? His garden today was alive with them. Looking at Washington's trees brought back a little humanity to a time that can seem cold and distant, and was filled with heartbreak for so many.

I toured Washington's home. From the beginning, I realized that the interpretation of this old southern plantation had shifted in the last few years. The past lives and work of the enslaved servants were foregrounded against the backdrop of idolatry. Caroline Branham, an enslaved housemaid, once carried a candle through a shadowy, sea-green New Room as she worked during those dim hours before sunrise. Christopher Sheels, serving as valet, once walked across the thick, heart pine floorboards of Washington's study. Doll and Hercules once worked in the hot kitchen, where temperatures soared to 120°F, preparing Washington's meals. All were slated to be freed in Washington's will upon the death of his wife.

After the tour, I felt empty still and headed back to my car. I drove past degraded land once planted with wheat, Washington's cash crop. I made my way down the gently curving road to Washington's old gristmill and distillery. I parked and walked toward the largest building on the property—at least three stories high, not including the attic space under the steeply gabled roof. There, I lost myself in the spinning of granite millstones and the whirring of sugar maple cogs. I was mesmerized by the fine cornmeal sifting through bolting cloth. At the distillery, I imagined enslaved distillers—Daniel, Hanson, James, Peter, and Timothy—heating up two-hundred-gallon boilers, transferring viscous liquid to mash tubs, cooking the rye and corn and malted barley, moving it to the stills. Those enslaved men worked under Scotsman James Anderson to produce more whisky than any other American distillery for the first president of the United States.

When I left the distillery building, the landscape grounded me again: it was green, but so barren. I saw few signs of animal life and barely heard birdsong. I made my way to the gift shop, wanting to feel strong again, and I gravitated to a yellow pair of socks with a black silhouette of a coiled rattlesnake on them. "Don't tread on me," they said. And I thought, yes, that's right. Don't tread on snakes. Don't tread on animals. Don't tread on Mother Nature. And I impulsively bought those socks.

When I had arrived in Washington, DC, ready to enjoy a visit with my closest friend, I left my new rattlesnake socks on the center console next to the driver's seat of the car. After visiting for a few hours at her house, my friend and I left to meet her husband for dinner at one of my favorite restaurants in the DC area. We decided to take my car.

Immediately after getting settled in the car and buckling her seatbelt, my friend noticed the bright-yellow socks emblazoned with a rattlesnake. She visibly flinched, and her head gave a tiny shake, "Nicki, what is this?" I knew something was wrong, and I explained that I had been feeling low and I bought some snake socks. "Nicki, no. You know what these represent, right? You can't wear these."

Now here's where the world seemed to invert a bit for me. I was familiar with "don't tread on me" as a symbol of the American Revolution. I had become obsessed with the American Enlightenment at age eleven. I also knew that, for some folks in the mid-1800s, the rattlesnake felt like the rightful symbol of the country, not a big-headed eagle. I also recognized having seen some people with "don't tread on me" bumper stickers over the last several years, and I wondered if they, like me, had an interest in the Revolutionary War or if they were just trying to reembrace the toughness of character that symbol represents. I was ignorant of the extent to which this early colonial symbol had been appropriated by modern politics.

Rob Walker, reporting in the *New Yorker,* explains how the symbolism of the insignia on what is known as the Gadsden flag has shifted since its inception during the colonial era. In 1751, Benjamin Franklin started to use the rattlesnake as an anti-British and pro-American Revolution symbol. He even published a political cartoon that featured a segmented snake, labeled with the American colonies' names, emblazoned with "Join or Die." Other colonial revolutionaries picked up on this snake imagery—including Paul Revere and Christopher Gadsden, from South Carolina. According to Walker, and a U.S. government directive that he cites, the Gadsden flag was purely a symbol of the Revolution. The Georgian twenty-dollar bill, released in 1778, might explain the symbol's meaning best with the Latin motto *Nemo me impune lacesset,* or "No one will provoke me with impunity."

That changed in 2001, when a Libertarian started a website on the history of the Gadsden flag. At that time, the flag experienced a resurgence, especially as the Tea Party movement ramped up. The Gadsden flag is now a specialty license plate option in a number of southeastern and western states. More recently, however, the flag has become associated with white supremacy. Just as Ku Klux Klan members fly the Confederate battle flag, white supremacists have started to use the Gadsden flag. And while it was high time that this Pollyannaish idealist got her head out of the sand, I realized I wasn't the only one uncertain about the meaning of this symbol today. Some think that the Gadsden flag, particularly on cars, means "don't tailgate"; others still firmly associate the flag with "those damn redcoats," seeing it as a homage to history. As Walker notes, symbolism evolves.

Snakes are largely absent from daily images in Western society, although two are intertwined on the Caduceus representing commerce, and one can still be seen singly on the Rod of Asclepius representing health care. However, these symbolic snakes are generally assumed to represent a nonvenomous colubrid from Europe, the Aesculapian Snake, which was introduced to temple sites devoted to healing around the ancient Greek world and were incorporated into rituals for wellness. This means that venomous snakes are largely absent from any positive symbols.

The Gadsden flag was different. The Gadsden flag showcased the Timber Rattlesnake, a venomous, bulky ambush predator that occupied forests in all thirteen original colonies with a range extending past the Mississippi River and into the American West. The Timber Rattlesnake, like many venomous snakes in the United States, packs a potent punch of toxins when it delivers a bite, but it typically needs to be provoked to do so. As was written in 1775, the Timber Rattlesnake's bite is "decisive and fatal," but "she never wounds 'till she has generously given notice." That was part of the message that the American colonists were sending to Britain, but it wasn't all of it.

Benjamin Franklin, writing under the pen name "An American Guesser," best captured the "worthy properties" of the Timber Rattlesnake and applied that symbolism to the American character. Franklin noted the ancient association of snakes with wisdom. He saw the Timber Rattlesnake's bright and always open eyes as "an emblem of vigilance." Moreover, Franklin contrived a situation where he had observed that the Timber Rattlesnake had exactly thirteen segments on its rattle, and noted with significance that "this was the only part of the Snake which increased in numbers." He also observed that Timber Rattlesnakes are independent most of the year but hibernate together to survive, extolling the virtue of cooperation to preserve the lives of the colonists. Perhaps most touching, Franklin pronounced the Timber Rattlesnake "beautiful in youth and her beauty increasth with her age," and thus established a hope for America, too.

While some of the ideas of the Revolution were beautiful, with the early colonial pronouncements that "all men are created equal" reverberating across the globe, America has not lived up that ideal. Upon gaining control of the colonies, the American colonists continued to drive Native Americans from their homelands and enslave Black Americans. Moreover, the American enterprise of expansion and resource exploitation for economic gain damaged and degraded the rich natural heritage of the country, including the very symbols of the United States' founding—the Bald Eagle and the Timber Rattlesnake.

Once found in all thirteen original colonies, today the Timber Rattlesnake has been extirpated completely from Rhode Island. Moreover, it's listed as endangered in New Hampshire, Massachusetts, Connecticut, New York, New Jersey, and Virginia. Moving beyond the colonial boundaries of the United States, the Timber Rattlesnake has also been extirpated from Maine and is endangered in Vermont, Ohio, Indiana, Illinois, Minnesota, and Texas, as well as Canada.

A *JUBO* IN CUBA

ONE WINTRY DAY IN the Piedmont, rainy and about 40°F, I was sitting on a big boulder in the middle of the Eno River, an ancient waterway, semi-protected but still littered by a pink balloon, its red string, and a gas meter flyer. I watched the water moving away, quicker and more turbid than normal. I watched the raindrops hit the surface of the river, leaving dimples and rings and so many circles. My thoughts went with those raindrops down the Eno River into the bigger and mightier Neuse, emptying into Pamlico Sound and into the Atlantic, getting swept up into the cerulean waters of the Gulf Stream, heading toward Europe, breaking off to the west to join the Azores Current, and then heading down south on the Canary Current to meet the Atlantic North Equatorial Current, and finally, hitting the Antilles Current to bring me where I really wanted to be: Cuba.

I was dreaming of going back to Cuba, dreaming of seeing a Broad-banded Dwarf Boa or a tiny "night stalker" snake in the genus

Arrhyton or a Majá de Santa María, Cuba's largest and most famous endemic snake. Cuba possesses a peculiar biological richness in reptiles: of its 140+ reptile species, 80 percent are found nowhere else. Cuba also boasts at least thirty species of snakes, though relatively little is known about their natural history. Cuban herpetologists suspect that new snake species could still be discovered on the island. While Cuban herpetofauna is the stuff of daydreams, I was also missing my Cuban friends.

One of my friends, Dany, is probably the most good-hearted, considerate man I know. He used to work as a firefighter in Havana. He trained other firefighters to drive those big 7,500- and 15,000-kg firetrucks, he worked hard and became a lieutenant, and he made thirty bucks a month. You can't support a family on thirty dollars a month, not in Cuba. So Dany left the job that he was exceptionally well-suited for to become the driver of a big beautiful old Cuban taxi, a 1955 Pontiac Chieftain, that tourists adore. That's how we met. I was one of those tourists, back in the Obama days, when Cuba was opening up to Americans again. I ended up returning to Cuba three more times within a year, and I was fortunate enough to become the unofficial sister of two remarkable brothers, Dany and Magdi.

Now Dany knows I love snakes, and as considerate as he is, I think he would like to like them just to please me. But he doesn't. This brave firefighter—a guy who has rushed into burning buildings, fought off three men at one time, and nearly had his finger bitten off by an aggressive Cuban parrot that he couldn't help but tease—was pretty freaked out by snakes.

It took him a while to tell me why. When Dany was twelve years old, walking home from school along a dirt road outside of Santa Clara, he found a small snake, a *Jubo de Sabana* or *Jubo Común* or Cuban racer (*Cubophis cantherigerus*). Dany, being naturally curious and an animal lover, approached the *jubo,* but it raised up its head and attacked him with surprising speed. Dany related that the *jubo* hit him with its tail, on the back of Dany's tan calves and on top of his leathery, bare foot, chasing him for a meter or two. Whether it was a tail or teeth, something hit Dany, because he was left with red lesions on both legs. He was very scared, Dany said, and after that he didn't want to meet a snake ever again.

I remember that when I first heard Dany's story about the *jubo*, I was skeptical. I'm always skeptical when someone tells me they or someone they know was attacked by a snake. It brings to mind old stories of "hoop snakes" biting their own tails and forming a hoop and rolling to attack people or of milk snakes found suckling from the cow in a barn in the central United States. Folklore is fun but often ludicrous. So, I imagined that Dany's fear and the time that had passed since his incident with the snake perhaps had clouded his memory, but it turns out that *Jubos de Sabana* do have a reputation. These long racers, with blue-gray bodies, hang out wherever they can survive— trash dumps, under rocks, in tall grass. They are also recorded to act like African cobras when threatened, raising the front of their body and expanding their neck.

Plus, this is a quick-moving, rear-fanged snake that can pack a bit of a punch, like a bee sting, maybe worse. In 1980, researchers described the toxic enzymes in the *jubo*'s saliva as being powerful enough to leave "bothersome lesions." In 1954, one herpetologist, Wilfred T. Neill, decided to test whether the *Jubo de Sabana* did indeed have a mild venom by letting a large adult bite him on his forearm and chew for two minutes. Within four minutes, the bite wound began to swell. Within ten minutes, the wound was puffy and quite red. Within forty-five minutes, the skin around the wound became dark and discolored, and streaks of red shot up two inches from the bite mark. Within four hours, the wound was a bruise two inches around. According to the author, the bite stayed discolored and painful for three more days. This detailed description led me to suspect that Dany was actually bitten by the *jubo*, and it was the streaks that made him think that perhaps it had whipped him with its tail, but I wasn't there.

In some cases, the effects of being bitten by a *jubo* can be much worse. One man, Crescencio Moreira, ended up in the hospital for several days in 1980 after being bitten on the thumb by a *jubo*. The inflammation extended well past his hand and arm, and even encompassed part of his chest. This wasn't the first time such a severe reaction to a *jubo* bite had been documented. In 1873, don Felipe Poey described a nearly identical case with swelling extending to the chest in only six hours. The patient was paralyzed in the affected areas and the effects lasted six full days. Even after a year and a half, the patient's fingers still weren't moving quite as they did before. The 1873 account also included a warning to mothers to not let their kids torment dogs because, like the innocent *jubo*, perhaps their harmless teeth could grow venomous when bothered.

In fact, the *Jubo de Sabana* was the first and only snake I ever saw in Cuba. It wasn't until the fourth trip, on a cool December day, with the sun shining bright that I saw it, or, rather, heard it. We had been walking across this fascinating karst landscape, sandstone on top of limestone, eaten away by water and time. Every footfall was uneven and bumpy, every step might land in a hole leading to a massive Cuban cave, so we walked carefully and quietly, which meant that I heard the snake first, that slow dragging sound, unhurried, relaxed. And there it was . . . a racer with a dark patch on the forehead and black "eyelashes" along the neck. I'm not sure it ever knew we were there.

It was after seeing that *jubo* that my husband said to me, "You know, the old me would have immediately picked that up." Mmm . . . hmmm . . . yep, the old him would have. "But I'm glad we just let them be now." And in that moment I realized that maybe an approach to herpetology, to wildlife observation that was truly observational might have some value to others, in a kind of Zen-like meditative way. Maybe the world could become kinder by showing respect even to the *Jubo Común*, who on some days might flatten its neck and give chase and on others might slither away happily in the sun. Maybe the world could become kinder by remembering the gentle words of awe embodied in "Serpiente," meaning snake, a poem written by Havana native Dulce Maria Loynaz (1902–1997):

It is made of the rings of Saturn,
of the moisture from the well, of the light of will-o'-the-wisp.
It signifies infinity if it bites its tail
and invites questions with its body raised.

His electric eye shines in the grass
and a sweet shiver unscrews it.

HOPE FOR THE FUTURE

 ON A COOL, CLEAR, late-March morning, we passed a small Baptist church on the left and drove a little way down a hill until we found a cinnamon-hued gate and a rutted, dirt road. A creek ran alongside the road, lined with Swamp Azaleas and cane, fringed with tall, thick trees, Swamp Tupelos and Sweetgums, Red Maples and Water Oaks. The other side of the dirt road was bordered by upland forest. Southern oaks stretched to the sky, still bare limbed. A lone Loblolly Pine, with its blocky auburn bark and straight, green needles sat nestled among a thicket of prickly greenbrier vines and shoulder-high viburnum bushes.

We pulled up to a cabin, an efficient and well-equipped two-bedroom house covered in thin slats of taupe siding, windows and doors rimmed in white, and adorned with two perfect porches, one screened in and furnished with a table and comfortable chairs, the other open to the air and bedecked with the tools of the trade: dip nets and snake hooks, knee-high Wellingtons and rugged boots, a terrarium and a birdhouse. We had arrived at Salleyland, the one-hundred-acre South Carolina property of the renowned herpetologist Whit Gibbons.

My husband, son, then ten years old, and I jumped out of our blue Honda CR-V, filled to bursting with waders and camping gear. We were greeted by the Sage of Salleyland himself, dressed in blue jeans and a black shirt, layered with a nice cambric and an olive-and-black-checked flannel. At eighty years old, Whit could have passed for sixty-five. His full head of white hair brushed the tops of his ears and covered his wide forehead. His blue eyes sparkled and he smiled wide as he introduced himself to my husband and son and gave me a hug.

Whit and I had met nine months earlier at a small herpetology gathering at the Groton School in Massachusetts. There we listened to presentations about the pros and cons of citizen science apps used to collect data on herps, like HerpMapper and iNaturalist, searched for ways to connect college and high school environmental curricula, and explored the Massachusetts woods. We found some New England herps that clear-skied June day: a fingerling-sized Common Garter Snake, Spring Peepers and Wood Frogs, and Red-spotted Newts and Red-backed Salamanders.

This time, we were searching for the salamanders of the Southeast, and I was bringing a group of Duke University graduate students interested in herpetology, four talented women ranging in age from twenty-four to thirty-one, and two of their significant others, one of whom was a gifted herpetologist himself studying at Auburn. The students were due to arrive at noon, and we had a few hours to pass. We spent them, of course, herping.

Donning rubber boots and grabbing snake hooks, we walked back down the gravel road to meet Whit's sixteen-year-old grandson, Parker. Parker had longish sandy-blond hair and the same wide smile and bright-blue eyes of his grandfather. He was friendly and agile and dedicated, wearing a royal-blue shirt that announced, "I'd rather be herping." Parker was also quite experienced; today he was looking for his one-hundredth herp species, and Whit, who wasn't prone to bragging, suspected that Parker was the most accomplished young herpetologist in the United States. It was impossible to disagree; Parker's quick hands could capture lizards in midair, and his quick mind could identify them too, along with other taxa.

And then we were off. With the music of White-eyed Vireos, Carolina Chickadees, and Northern Cardinals serenading our adventure, we began flipping cover boards and setting out minnow traps. It wasn't long before we found salamanders: gold-and-black Three-lined Salamanders, white speckled Slimy Salamanders, and brilliant orange-red Mud Salamanders.

We walked up to a patch of sphagnum moss, bounded by Red Bays and Swamp Tupelos. Parker pulled up the moss in clumps and set the clumps back in place rapidly until, finally, he got it: a Chamberlain's Dwarf Salamander. A diminutive salley, stretching no more than two inches long, its belly bright yellow and its tan back lined with three broken black stripes. This was a life-lister for me.

While we rummaged for herps, I gained a fuller picture of Whit. This was a man who had published at least twenty books and hundreds of articles in the field of herpetology. A man who spent years at the famous Savannah River Ecology Lab and was a professor at the University of Georgia. Deservedly, he had won numerous awards in herpetology, including the IUCN Behler Turtle Conservation Award and the Henry Fitch Distinguished Herpetologist Award, and more awards honoring his work as an environmental educator. With all those accomplishments, Whit could have been pedantic and arrogant, but he wasn't. Instead, he was quick to laugh, interested in others, and careful to boost people up rather than bring them down.

As the students arrived, Whit conscientiously recited their names. He knew them all, despite never having met any of them. He welcomed everyone graciously, helped them get settled in bunks in the cabin or in tents nearby. And then he began to get to know everyone—inquiring after their hidden talents, finding their areas of expertise, and noting their interests. Throughout the trip, Whit would quietly call me over and say something nice about each student, things like, *she did an excellent job communicating,* or *if I were still a professor, I'd want these students in my lab.* He did the same for me. Every once in a while, the students would find a salamander or a tadpole and bring it to Whit to identify, understandable as he had literally written the book on every herp taxa in the southeastern United States. But instead of identifying it or ploddingly detailing every last fact and characteristic of the species, he'd say something like, *go ask Dr. Cagle.* I had never been in an environment before where an older, wiser professor would so magnanimously and conscientiously raise up a younger, less-learned colleague in that way.

That first day was a herpetological extravaganza, especially for those of us still in the midst of the tail end of winter in North Carolina. In a few hours, we had racked up seventeen species of herps, scurrying Ground Skinks and blue-bellied and virile Fence Lizards, slider turtles and Southern Leopard Frogs, and, of course, a couple of Ringneck Snakes that slid between our fingers like smooth strings of black pearls. And yes, Parker did get his one-hundredth herp that day: the strange Many-lined Salamander with its finely inked back and flattened tail.

That evening, the student from Auburn, Philip, made an incredible fire. He piled tinder, then small sticks, then large sticks, then arm-thick branches into the fire pit until the flames shot five feet into the air. The tall flames soon settled into a warm and welcoming campfire, conducive to conviviality on a cold evening. Whit pulled out a couple bottles of wine for us and poured himself a bit of gin. We sat around the fire, situated between the cabin and the swamp. We ate and drank and talked.

The conversation went in many directions. Whit answered questions about his life. Both of his parents had taught at the university level: his mother had studied English but taught marketing; his father had studied English and wrote at least two novels. Harper Lee, of *To*

Kill a Mockingbird fame, would visit their home and even gleaned material from one of Whit's own childhood experiences. Listening to Whit, as he calmly sipped gin or puffed on a Boneshaker cigar, as his eyes twinkled with interest or his shoulder bumped mine in comradery, I began to see that Whit belonged to another generation. He belonged to the tail end of the Silent Generation, a generation that was usually so tight-lipped about their experiences that they were only knowable through old movies and the relics of their past. Whit was different: he was sharing, and in sharing he showed deep respect for our small, mixed-generational group representing the Oregon Trail Generation, Millennials, and Generation Z.

Whit showed respect to this intergenerational group in a way that was perhaps more profound: he asked questions. He asked me about the movement embracing pronouns (mine are she, her, hers). He listened carefully to the answer and requested that I send him an article I had mentioned on the subject. He asked about #Me Too. He listened carefully as a student explained how men always were asking her why she didn't smile or telling her that she should. He asked us all to share our first memories of a world event. The Berlin Wall coming down. The oil fields burning during the first Iraq War. Columbine. The planes crashing on September 11. The inauguration of Barack Obama. Whit's own first memory was listening to the radio when Franklin Delano Roosevelt died in April 1945. He also remembered standing six feet away from Governor George Wallace as he proclaimed that he would preserve "individual freedoms of citizens" by forbidding integration in Alabama public schools in June 1963. He remembered John F. Kennedy being assassinated in November later that same year. Whit leaned into me saying that he liked being old because he had all his memories and shared the memories of the students as well.

Michael Meade, an author and mythologist, describes the difference between an "older" and an "elder." He says that in our society we have many olders, people who have lived long but have not gained wisdom. He also suggests that what we need is elders, people who have "grown deeper" and become wiser. Like Whit Gibbons, elders make a shift from living shallowly to living deeply, from living meaning to do something to living with meaning, from living to serve their own ego to living to serve others.

Elders are key figures in traditional cultures, guiding individuals, helping the larger community, and speaking for the natural world. By contrast, modern American culture is largely bereft of elders, and it shows. In every aspect of our lives, people are desperately working to develop cultures of eldership and mentorship, turning to people outside of our own communities to remind us of the qualities required to be a good elder: knowledge, wisdom, compassion, a willingness to take on responsibility, and a call to teach. There are eldership workshops in Italy, mentoring events at midwestern business development centers, and mentor trainings at Big 10 universities. There are entire schools of environmental education built around mentoring, like Jon Young's Coyote Way.

The point is, we need more Whit Gibbonses. We need more herpetologists speaking for the reptiles and amphibians, more botanists speaking for the plants, more ecologists speaking for the earth. This "speaking for" doesn't make our observations any less objective or our science any less rigorous. Herpetologists, botanists, and ecologists have always been people, people with fascinations, people with perspectives, people with opinions. Rather than detracting from our work, speaking for what we know best marries facts with a responsibility to those facts. That union has a name: wisdom.

With their deep knowledge, scientists must also fill the role of elders. We need more scientists like Whit Gibbons. Someone who will invite children, young adults, undergraduate and graduate students, young and midcareer professors to an oasis in rural South Carolina. Someone who will guide them with a light touch as they explore the nooks and crannies of sphagnum wetlands and upland dry forests. Someone who will share their knowledge selectively and defer to others out of a sense of deep generosity. Someone who lifts up the next generation as they learn about the world and grapple with the difficulties it presents.

The next generation of herpetologists and conservationists seemed to be having a rather good time on this visit to Salleyland. After a late night chatting around the fire, the students woke up, gulped down some coffee, and were ready to go back out into the field. The wet woods of the morning were filled with birdsong—the rasping of Blue-gray Gnatcatchers, the chirrups of towhees. The students were snapping photos and bringing them to me to confirm identification. *Is this a goldfinch?* Yes, it's not yet in full breeding color. *Is this a Song Sparrow?* Maybe. It's kinda small. Was it that red in real life?

After satisfying ourselves with the bottomland birds, we decided to check the traps we had set out in the creek and wetlands: trash can traps with funnels, minnow traps, and hoop traps. Everyone spread out along the trails, pulling up dripping traps, feeling the buzz of excitement as they peeked inside. The traps yielded mud minnows and fliers, pickerel and bluegills, and even a Pirate Perch, with its strange, migrating anus. The Pirate Perch is small, stout, and coffee-hued. It is also the only species in its family. When the young hatch, their anus sits near their anal fin, the position most of us would expect. But as the fish grows, the anus slowly moves forward, ending up in line with the pectoral, or front, fin. This I learned from Whit Gibbons's knowledgeable grandson, Parker. This strange change results in an even stranger scientific name, *Aphredoderus,* which translates to "excrement throat."

We also found bright-red crayfish with enormous claws and small, crabby stinkpot turtles. The stinkpot, also known as the Eastern Musk Turtle, was found exactly where one would expect: in ponds of the lower Piedmont with soft, muddy bottoms. While rather plain, stinkpots do have a neat adaptation. Young stinkpots are especially known for the foul scent produced from four glands that release stinky, yellow liquid. This musky odor likely warns away predators and might assist with homing, courtship, or other interactions within the species. Some of these have been tested in labs, where researchers found that male stinkpots tend to be attracted to water scented by females.

The search continued. Some students pulled up traps splashing with Bronze Frog tadpoles with their dark, red-tinged bodies, nearly translucent dorsal fin and solid tail musculature, both dusted with dark speckles. Another group pulled out an American Bullfrog. My son began pulling back clumps of bright-green sphagnum moss, uncovering another Chamberlain's Dwarf Salamander to the delight of those who hadn't seen the first.

The morning disappeared into midday. Birds still called, only now we heard the nasal caws of the Fish Crow and the bubbling beeps of Blue Jays. We admired the delicate pink blooms of the Swamp Azalea and bright samaras of the Red Maples. But something was missing from our list. We all began flipping cover with calm determination. We flipped the first board. Nothing. The second. Nothing. The third. Ants. We flipped tin. The first tin. Nothing. The second and third. Nothing. The tenth. Nothing. Until finally, one of the graduate students—donning full chest waders, a tie-dyed blue shirt, and a knit cap—lifted a large tin. "Snake!" she yelled. My husband ran over, dropping to his knees to grab it while Kelly held the corrugated metal high.

It was a Southern Black Racer, sleek and calm. Besides a little white under its chin, it was jet black. It stretched about three feet long and had a small bulge about a third of the way down its body, where a meal was still being digested. Black Racers always seem aware and intelligent. Their big, black eyes are prominent on their small face. As Whit suggests, these snakes clearly use their eyes for hunting.

Everyone was excited. For herp lovers, no day feels complete without finding a snake. We passed around the quiet, knowing Black Racer. Kelly's face lit up in a wide smile when she held it, and she cocked her head to the side when I snapped a photo. Philip, the fire starter from Auburn, dressed in his blue flannel and olive vest, held the snake with one hand, confident and self-possessed.

I saw hope for the future. A new generation of herpetologists and conservationists. A generation with men and women and gender-nonconforming people tramping around the fields and forests as equals. A generation where showmanship and daring were replaced with respect and awe. A generation tasked with conserving some of the most rapidly disappearing taxa on earth.

Nick's turn to hold the racer came next. He was a Duke graduate student's partner, and his world revolved around supporting autistic schoolchildren. He held the snake differently, letting it intertwine between the fingers of both his hands. He watched the sleek Black Racer with awe and wonder.

ACKNOWLEDGMENTS

The tapestry of our interconnectedness cannot be unwoven, and John Muir, flawed as he may have been, got it right when he said, "When we try to pick out anything by itself, we find it hitched to everything else in the Universe." When writing a book from a place of passion, as this one was written, the list of those for whom I feel grateful, of those who deserve acknowledgment, is endless. This book represents a (half) lifetime of development and experience, and thus it feels proper to acknowledge everyone I have met on this journey. Without a single one of the people who have touched my life, this book wouldn't be what it is.

And yet, of course, I do want to acknowledge that in any tapestry, there are some strings that frame the overall piece, those lengthwise warp yarns that bear the weight of the meandering, over-and-undering weft. I am deeply grateful to my family and the natural world, the warp strings of my own life. My parents—Eileen and Joe Flocca—are the most supportive that anyone could have, and they introduced me to the natural world at such a young age that it has always been a part of me. My husband, Mark Cagle, has been a constant companion. His own deep love of herpetofauna and undeniable expertise have deepened our own adventures and saved and improved so many animal lives. My son, Grant, is an incredible soul—deeply curious and kind—and he is a constant guide for me, giving me a ruler by which to measure the rightness of my life.

Thank you to my friend-sisters. Claire O'Dea, you have been one of the most positive forces in my entire life, and I'm so thankful to have met you all those years ago on Duke's campus. Sara Childs, thank you for all of our long walks in the woods; I'd be lost without them. Noëlle Wyman-Roth, thank you so much for your conversation and amazing food. Thank you also to Julie DeMeester and Kat O'Brien and Carrie Seltzer for being both warp and weft in my life. Simon and Laura Woodrup, Kayleigh Somers and Matt Poland—I value every meeting. Lucas and Jamie Joppa—you are both so inspiring. Thank you to those old friends from our academy days—Jon Marron, Tony Ham, Kristin Chow, Matt Rosen, and Raj Patel. Thank you, too, to those from the old days who have made such an impact on who I am today—April Faith-Slaker, Ben Cohen, and Kevin Donnelly. Thank you to Magdiel Pérez Martínez and Jessica Tamayo Aguilera, Daynier León Martínez and Yenisey Tamayo García—your friendship has been *un regalo maravilloso*.

I also want to thank my fellow herp-lovers: Steve Swanson and all the dedicated staff at The Grove who helped shape my way, especially Lorin Ottlinger, Emily Loeks, Kathryn Elble, Peter Babikan, and Patti K; Dave Holtzman and the Ometepe gang—Teresa Cochran Alvarez, Karen Brenny, Dave Steen (who is doing amazing work, check out *Secrets of Snakes*), Byron Jackson Holcomb, Ira, Melissa, Jaime Matos, and Matt Helmus, all of whom I first had the pleasure of meeting all those years ago in Costa Rica; Ron Sutherland and Ron Grunwald, who share my passion for local herpetofauna and have been so generous with their own knowledge; and students who have shared their passion for herps—Caitlyn Cooper, Hannah Royal, Diego Calderon-Arrieta, Julia Geschke, Kendra Sultzer, Madison Cole, Troi Perkins, Kelly Joyner, and those students who have encouraged me to share mine, like Gabby Buria, Mike Asch, Katie Myers, Erika Reiter, and Lannette Rangel.

And thank you to all my teachers, academic and beyond: Mrs. Di-Vito, Mrs. Hines, Mrs. Miller in seventh grade for encouraging me to write and taking me and Leslie Tone to the *Chicago Tribune;* Mr. Jozwik, Mr. Blackwell, and Hilary Rosenthal for teaching me how to think critically; Señora Julia Guerrero, who left this world too soon and gave me the travel bug that changed the course of my life; Chris Hilvert and Jeff Yordy (also gone too soon) for making science cool and never making me feel out of place as a girl in science; Peggy Whalen-Levitt and the Center for Education, Imagination and the Natural World; Dean Urban, Norm Christensen, John Terborgh, and Dan Richter for your guidance and continued support. I am also grateful for the support of my colleagues at Duke—particularly Sari Palmroth, Rebecca Vidra, Liz Shapiro-Garza, Charlotte Clark, Liz DeMattia, Meagan Dunphy-Daly, Martin Doyle, Toddi Steelman, Joe Bachman and Paul James (we miss you both!), Nico Cassar, Jeff Vincent, Lori Bennear, John Poulsen (who co-advised Hannah Royal and Kelly Joyner), Blake Tedder, Beverly Burgess, Allison Besch, Sandi MacLachlan, Danielle Wiggins, Nancy Kelly, Melissa Kotacka, and Tom Brooks—thank you for your encouragement and consistent generosity of spirit.

Many thanks, too, to Angie Hogan at the University of Virginia Press, for believing in this book, Susan Murray, and to the three reviewers, including David Steen, who provided such thoughtful and good-hearted comments. You have made this work better. Thank you, Whit, gumby, Diego, Hannah, Kelly, Nick, and Lannette for reviewing your chapters and sharpening my memory and prose. Thanks, again, to Claire, my parents, Mark, and Grant (you are an amazing editor!) for their early feedback on this manuscript.

I also want to extend deep gratitude to my extended family—all of you in Illinois and now spread out across the country including the Azzanos and Malinowskis, the Kukulskis and Bergers—I have a wonderfully supportive network of aunts, uncles, and cousins; my *famiglia* Flocca—Giusi *e i tuoi figli, così come* Nina, Pietro, and Francisco in Sicily and other family in the rest of Europe, all those who I have met on my travels, and to my ancestors. You are all part of this tapestry.

REFERENCES

AN ACRE OF SNAKES

Cagle, Nicolette L. 2008. "Snake Species Distributions and Temperate Grass-lands: A Case Study from the American Tallgrass Prairie." *Biological Conservation* 141 (3): 744–55. https://doi.org/10.1016/j.biocon.2008.01.003.

Dalrymple, G. H., and N. G. Reichenbach. 1984. "Management of an Endangered Species of Snake in Ohio, U.S.A." *Biological Conservation* 30: 195–200.

David, Gary A. 2018a. "Dances with Snakes: The Real Reason for the Hopi Snake Dance." April 9, 2018. https://www.ancient-origins.net/history -ancient-traditions/dances-snakes-real-reason-hopi-snake-dance -009868.

——. 2018b. "Dances with Snakes: The Real Reason for the Hopi Snake Dance—Part II." April 10, 2018. https://www.ancient-origins.net/history -ancient-traditions/dances-snakes-real-reason-hopi-snake-dance-part-ii -009870.

Ernst, C. H., and R. W. Barbour. 1989. *Snakes of Eastern North America.* Fairfax, VA: George Mason University Press.

Most, Matthew G. 2013. "Activity Patterns and Spatial Resource Selection of the Eastern Garter Snake (Thamnophis sirtalis sirtalis)." Master's thesis, Loyola University Chicago.

Myers, J. Jay. 2018. "The Sacred Hopi Snake Dance Impressed Theodore Roosevelt." May 10, 2018. https://www.historynet.com/sacred-hopi -snake-dance-impressed-theodore-roosevelt.htm.

Nilsson, Martin P. 1940. "Greek Popular Religion: The House and the Family." https://www.sacred-texts.com/cla/gpr/gpr08.htm.

Reading, C. J., L. M. Luiselli, G. C. Akani, X. Bonnet, G. Amori, J. M. Bal-
louard, E. Filippi, G. Naulleau, D. Pearson, and L. Rugiero. 2010. "Are
Snake Populations in Widespread Decline?" *Biology Letters* 6 (6): 777–80.
https://doi.org/10.1098/rsbl.2010.0373.D.

LESSONS FROM WISCONSIN

Hotta, Eri, Risa Tamagawa-Mineoka, Koji Masuda, Maiko Taura, Yuka
Nakagawa, Fuminao Kanehisa, Saki Tashima, and Norito Katoh. 2016.
"Anaphylaxis Caused by γ-Cyclodextrin in Sugammadex." *Allergology
International* 65 (3): 356–58. https://doi.org/10.1016/j.alit.2016.02.013.
Ranayhossaini, Daniel J. 2010. "An Investigation of the Hemotoxicity of
the Duvernoy's Gland Secretion of the Northern Water Snake (Nerodia
Sipedon)." Baccalaureate thesis, Pennsylvania State University, Schreyer
Honors College. https://honors.libraries.psu.edu/files/final_submissions
/272.
Steen, David A. 2019. *Secrets of Snakes: The Science beyond the Myths.* College
Station: Texas A&M University Press.

IDOLS

Allender, Matthew C., Jennifer A. Moore, Eric T. Hileman, and Sasha J.
Tetzlaff. 2015. "Ophidiomyces Detection in the Eastern Massasauga in
Michigan." University of Illinois Wildlife Epidemiology Laboratory.
Bernstein, Richard. 1996. "The Beauty of Life, Including Snakes." *New York
Times*, August 23, 1996. https://www.nytimes.com/1996/08/23/books
/the-beauty-of-life-including-snakes.html.
Cagle, Nicolette L. 2008. "A Multiscale Investigation of Snake Habitat Rela-
tionships and Snake Conservation in Illinois." Ph.D. diss, Duke University.
Cornett, Peggy. 2001. "Thomas Jefferson's 'Belles of the Day' at Monticello."
https://www.monticello.org/house-gardens/center-for-historic-plants
/twinleaf-journal-online/belles-of-the-day/.
Ellison, George. 2014. "Bryson City Nature Journal: Check out Peattie's Writing." *Citizen Times.* https://www.citizen-times.com/story/life/2014/02
/26/bryson-city-nature-journal-check-out-peatties-writing/5849603/.

Ellsworth, Nathaniel. 2020. "2020 Rattlesnake Roundup in Sweetwater Not Cancelled Because of Coronavirus." *Abilene (TX) Reporter-News.* https://www.reporternews.com/story/news/local/2020/03/12/rattlesnake-roundup-sweetwqternot-shaken-virus-worries/5033618002/.

Franke, Joseph. 2000. "Rattlesnake Roundups: Uncontrolled Wildlife Exploitation and the Rites of Spring." *Journal of Applied Animal Welfare Science* 3 (2): 151–60. https://doi.org/10.1207/S15327604JAWS0302_7.

The Humane Society of the United States. n.d. "Rattlesnake Roundup Chart." https://www.humanesociety.org/sites/default/files/docs/rattlesnake-roundup-chart.pdf.

Kennicott, Robert W. 1854. "Rattlesnakes." *Prairie Farmer*, December 15, 1854.

Keyes, Allison. 2017. "Two Smithsonian Scientists Retrace the Mysterious Circumstances of an 1866 Death and Change History." https://www.smithsonianmag.com/smithsonian-institution/two-smithsonian-scientists-retrace-mysterious-circumstances-1866-death-180962417/.

Kopp, Elizabeth Z. 2017. *The Kennicott Children: Carrying on the Kennicott Legacy.* Glenview, IL: Grove National Historic Landmark.

Lanesa, Nicoletta. 2019. "This Fungus Makes Snakes Look Like Mummies: It Just Turned Up in California." https://www.livescience.com/snake-fungal-disease-in-california.html.

Lillywhite, Harvey B. 2014. *How Snakes Work: Structure, Function, and Behavior of the World's Snakes.* New York: Oxford University Press.

Means, D. Bruce. 2009. "Effects of Rattlesnake Roundups on the Eastern Diamondback Rattlesnake (*Crotalus adamanteus*)." *Herpetological Conservation and Biology* 4 (2): 132–41.

Meyers, Mary Hockenberry. 2021. "What's in a Name?" https://grasstalk.wordpress.com/2021/11/22/whats-in-a-name/.

Peattie, Donald Culross. Ca. 1930. "A Natural History of Pearson's Falls and Some of Its Human Associations." http://toto.lib.unca.edu/booklets/natural_history_pearson's_falls/default_natural_history_pearson's_falls.htm.

———. 1935. "Suffering Snakes." *Saturday Review*, October 12.

———. 1941. *The Road of a Naturalist.* Cambridge, MA: Riverside.

———. 1991. *A Natural History of Trees of Eastern and Central North America.* Boston: Houghton Mifflin.

Quinn, Brother C. Edward. 1986. "A Zoologist's View of the Lewis and Clark

Expedition." *American Zoologist* 26 (2): 299–306. https://doi.org/10.1093/icb/26.2.299.

"Rattlesnake Roundup: A Texas Community Tradition." n.d. https://www.npr.org/sections/pictureshow/2020/04/03/821397097/rattlesnake-roundup-a-texas-community-tradition.

"Russia." Jefferson to John Adams. June 1, 1822. Coolidge Collection of Thomas Jefferson Manuscripts, Massachusetts Historical Society. Transcription available at Founders Online. https://founders.archives.gov/documents/Jefferson/98-01-02-2840.

Stanton, Lucia. 2013. "Thomas Jefferson and Virginia's Natural History." *Banisteria* 41: 5–16.

Steen, David A. 2019. *Secrets of Snakes: The Science beyond the Myths.* College Station: Texas A&M University Press.

"Sweetwater Jaycees." n.d. Sweetwater Jaycees. http://www.rattlesnakeroundup.net/.

Vasile, Ronald S. 1994. "The Early Career of Robert Kennicott, Illinois' Pioneering Naturalist." *Illinois Historical Journal* 87 (3): 150–70.

Walcheck, Kenneth C. 2008. "Montana Zoological Discoveries through the Eyes of Lewis and Clark." *We Proceeded On* 34 (3): 18–26.

Wang, Joyce Mujeh. 2008. "Rattlesnake Roundups." Unpublished student paper. Durham, NC: Duke University.

WikiVet English, s.v. "Snake Musculoskeletal System." n.d. Accessed February 20, 2020. https://en.wikivet.net/Snake_Musculoskeletal_System.

LOVE AND LOATHING

Friebohle, Jake, Shane Siers, and Chad Montgomery. 2020. "Acetaminophen as an Oral Toxicant for Invasive California Kingsnakes (Lampropeltis Californiae) on Gran Canaria, Canary Islands, Spain." *Management of Biological Invasions* 11 (1): 122–38. https://doi.org/10.3391/mbi.2020.11.1.09.

McRae, Mike. 2017. "Guam's Plague of Snakes Is Devastating the Whole Island Ecosystem, Even the Trees." https://www.sciencealert.com/guam-s-plague-of-snakes-is-having-a-devastating-impact-on-the-trees.

"Native Forest Birds of Guam." n.d. https://www.guampedia.com/a-native-forest-birds-of-guam/.

Öhman, Arne, and Susan Mineka. 2003. "The Malicious Serpent: Snakes as a Prototypical Stimulus for an Evolved Module of Fear." *Current Directions in Psychological Science* 12 (1): 5–9. https://doi.org/10.1111/1467-8721 .01211.

Rakison, David H. 2009. "Does Women's Greater Fear of Snakes and Spiders Originate in Infancy?" *Evolution and Human Behavior* 30 (6): 438–44. https://doi.org/10.1016/j.evolhumbehav.2009.06.002.

Savarie, Peter J., John A. Shivik, Gary C. White, Jerome C. Hurley, and Larry Clark. 2001. "Use of Acetaminophen for Large-Scale Control of Brown Treesnakes." *Journal of Wildlife Management* 65 (2): 356–65. https://doi .org/10.2307/3802916.

Torkar, G. 2015. "Pre-Service Teachers' Fear of Snakes, Conservation Attitudes, and Likelihood of Incorporating Animals into the Future Science Curriculum." *Journal of Baltic Science Education* 14 (3): 401–10.

Weber, Andreas. 2016. *The Biology of Wonder: Aliveness, Feeling, and the Metamorphosis of Science.* Gabrioloa Island, British Columbia: New Society.

LA SUERTE

Bolaños, Roger. 1984. *Serpientes, venenos y ofidismo en Centroamérica.* San José: Editorial Universidad de Costa Rica.

Lourenço-de-Moraes, Ricardo, Fernando Miranda Lansac-Toha, Leilane Talita Fatoreto Schwind, Rodrigo Leite Arrieira, Rafael Rogério Rosa, Levi Carina Terribile, Priscila Lemes, et al. 2019. "Climate Change Will Decrease the Range Size of Snake Species under Negligible Protection in the Brazilian Atlantic Forest Hotspot." *Scientific Reports* 9 (1): 1–14. https://doi.org/10.1038/s41598-019-44732-z.

John, Julia. 2017. "Reptile Atlas Highlights More Biodiversity Hotspots." https://wildlife.org/reptile-atlas-highlights-more-biodiversity-hotspots/.

Roll, Uri, Anat Feldman, Maria Novosolov, Allen Allison, Aaron M. Bauer, Rodolphe Bernard, Monika Böhm, et al. 2017. "The Global Distribution of Tetrapods Reveals a Need for Targeted Reptile Conservation." *Nature Ecology & Evolution* 1 (11): 1677–82. https://doi.org/10.1038/s41559-017 -0332-2.

Brown, Gregory P., and Patrick J. Weatherhead. 2000. "Thermal Ecology and Sexual Size Dimorphism in Northern Water Snakes, *Nerodia sipedon.*" *Ecological Monographs* 70 (2): 311.

Daerr, Elizabeth G. 1999. "Eastern Indigo Snake." *National Parks,* September 1999. Gale Academic OneFile.

Fitch, Henry S., and Richard A. Seigel. 1984. *Vertebrate Ecology and Systematics: A Tribute to Henry S. Fitch.* Lawrence: University of Kansas. https://doi.org/10.5962/bhl.title.5482.

Hanson, Britta A., Philip A. Frank, James W. Mertins, and Joseph L. Corn. 2007. "Tick Paralysis of a Snake Caused by Amblyomma rotundatum (Acari: Ixodidae)." *Journal of Medical Entomology* 44 (1): 155–57.

Kabay, Edward. 2013. "Timber Rattlesnakes May Reduce Incidence of Lyme Disease in the Northeastern United States." https://eco.confex.com/eco /2013/webprogram/Paper44305.html.

Kwak, Mackenzie L., Chi-Chien Kuo, and Ho-Tsung Chu. 2020. "First Record of the Sea Snake Tick Amblyomma nitidum Hirst and Hirst, 1910 (Acari: Ixodidae) from Taiwan." *Ticks and Tick-Borne Diseases* 11 (3): 101383. https://doi.org/10.1016/j.ttbdis.2020.101383.

Lambert, Helen, Gemma Carder, and Neil D'Cruze. 2019. "Given the Cold Shoulder: A Review of the Scientific Literature for Evidence of Reptile Sentience." *Animals* 9 (10): 821. https://doi.org/10.3390/ani9100821.

Macartney, J. Malcolm, Patrick T. Gregory, and Karl W. Larsen. 1988. "A Tabular Survey of Data on Movements and Home Ranges of Snakes." *Journal of Herpetology* 22 (1): 61–73. https://doi.org/10.2307/1564357.

Natusch, Daniel J. D., Jessica A. Lyons, Sylvain Dubey, and Richard Shine. 2018. "Ticks on Snakes: The Ecological Correlates of Ectoparasite Infection in Free-Ranging Snakes in Tropical Australia." *Austral Ecology* 43 (5): 534–46. https://doi.org/10.1111/aec.12590.

Reinert, Howard K., and Robert R. Rupert Jr. 1999. "Impacts of Translocation on Behavior and Survival of Timber Rattlesnakes, Crotalus horridus." *Journal of Herpetology* 33 (1): 45–61.

"Reptile Emotions." 2011. *News & Publications,* August 18, 2011. https:// vetmed.tamu.edu/news/pet-talk/reptile-emotions/.

Rock, Katelyn N., Isabelle N. Barnes, Michelle S. Deyski, Katheleen A. Glynn, Briana N. Milstead, Megan E. Rottenborn, Nathaniel S. Andre, Alex

Dekhtyar, Olga Dekhtyar, and Emily N. Taylor. 2021. "Quantifying the Gender Gap in Authorship in Herpetology." *Herpetologica* 77 (1): 1–13.

Stap, Don. 2001. "American Heritage—Tracking North America's Largest Snake." *National Wildlife* 39 (7). http://link.gale.com/apps/doc/A78578545/CPI?u=duke_perkins&sid=zotero&xid=ba87677f.

University of Maryland. 2013. "Timber Rattlesnakes Indirectly Benefit Human Health: Not-So-Horrid Top Predator Helps Check Lyme Disease." https://www.sciencedaily.com/releases/2013/08/130806091815.htm.

Webb, Jonathan K., Barry W. Brook, and Richard Shine. 2002. "What Makes a Species Vulnerable to Extinction? Comparative Life-History Traits of Two Sympatric Snakes." *Ecological Research* 17 (1): 59–67. https://doi.org/10.1046/j.1440-1703.2002.00463.x.

DISSERTATION

Anderson, R. C. 1991. "Illinois Prairies: A Historical Perspective." *Illinois Natural History Survey Bulletin* 34 (4): 384–91.

Baird, S. F., and C. F. Girard. 1852. In *Exploration and Survey of the Valley of the Great Salt Lake of Utah*, edited by H. Stansbury, 336–65. Philadelphia: Lippincott, Grambo.

———. 1853. *Catalogue of North American Reptiles in the Museum of the Smithsonian Institute*. Washington, DC: Smithsonian Institute.

Bavetz, M. 1993. "Geographic Variation, Distribution, and Status of Kirtland's Snake, Clonophis kirtlandii (Kennicott) in Illinois." Master's thesis, Southern Illinois University.

Brandehoff, Nicklaus, Cara F. Smith, Jennie A. Buchanan, Stephen P. Mackessy, and Caitlin F. Bonney. 2019. "First Reported Case of Thrombocytopenia from a Heterodon nasicus Envenomation." *Toxicon* 157 (January): 12–17. https://doi.org/10.1016/j.toxicon.2018.11.295.

"The Breadbaskets of the World." n.d. https://www.worldatlas.com/articles/the-breadbaskets-of-the-world.html.

Cagle, Nicolette L. 2008. "A Multiscale Investigation of Snake Habitat Relationships and Snake Conservation in Illinois." Ph.D. diss., Duke University.

Capparella, Angelo P., Todd Springer, and Lauren E. Brown. 2012. "Discovery of New Localities for the Threatened Kirtland's Snake (Clonophis kirtlandii) in Central Illinois." *Transactions of the Illinois State Academy of Science* 105 (3–4): 101.

Conant, R., and J. T. Collins. 1991. *A Field Guide to Reptiles and Amphibians: Eastern and Central North America*. Boston: Houghton Mifflin.

"Dr. Philo R. Hoy: A Racinian to Remember." n.d. https://www.bellecitymag .com/2019/04/27/dr-philo-r-hoy-a-racinian-to-remember/.

Edgren, R. A. 1952. "A Synopsis of the Snakes of the Genus Heterodon, with the Diagnosis of a New Race of Heterodon nasicus Baird and Girard." *Natural History Miscellanea Chicago Academy of Sciences* 112: 1–4.

Gallagher, Erin. n.d. "Midewin Prairie to Undergo Restoration." https:// www.chicagotribune.com/suburbs/daily-southtown/ct-sta-midewin -restoration-st-0819–20160818-story.html.

Garman, H. 1892. "A Synopsis of the Reptiles and Amphibians of Illinois." *Bulletin of the Illinois State Laboratory of Natural History* 3: 215–390.

Greenberg, J. 2002. *A Natural History of the Chicago Region*. Chicago: University of Chicago Press.

Harding, J. H. 1997. *Amphibians and Reptiles of the Great Lakes Region*. Ann Arbor: University of Michigan Press.

Herkert, J. R. 1991. "Prairie Birds of Illinois: Populations Response to Two Centuries of Habitat Change." *Our Living Heritage: The Biological Resources of Illinois* 34 (4): 393–99.

Hillis, David M. 2019. "Species Delimitation in Herpetology." *Journal of Herpetology* 53 (1): 3. https://doi.org/10.1670/18–123.

Hoekstra, Jonathan M., Timothy M. Boucher, Taylor H. Ricketts, and Carter Roberts. 2005. "Confronting a Biome Crisis: Global Disparities of Habitat Loss and Protection." *Ecology Letters* 8 (1): 23–29. https://doi.org/10.1111 /j.1461–0248.2004.00686.x.

Kennicott, R. 1856. "Description of a New Snake from Illinois—Regina kirtlandii." *Proceedings of the Academy of Natural Sciences of Philadelphia* 8: 95–96.

LeClere, Jeffrey B., Erica P. Hoaglund, Jim Scharosch, Christopher E. Smith, and Tony Gamble. 2012. "Two Naturally Occurring Intergeneric Hybrid Snakes (Pituophis catenifer sayi × Pantherophis vulpinus; Lampropeltini, Squamata) from the Midwestern United States." *Journal of Herpetology* 46 (2): 257–62. https://doi.org/10.1670/10–260.

McClain, W. E. 1997. *Prairie Establishment and Landscaping*. Natural Heritage Technical Publication no. 2. Illinois Department of Natural Resources.

Noss, R. F., E. T. LaRoe III, and J. M. Scott. 1995. *Endangered Ecosystems of the United States: A Preliminary Assessment of Loss and Degradation*. National Biological Service.

Phillips, C. A., R. A. Brandon, and E. O. Moll. 1999. *Field Guide to Amphibians and Reptiles of Illinois*. Champaign: Illinois Natural History Survey.

Robertson, K. R., R. C. Anderson, and M. W. Schwartz. 1997. "The Tallgrass Prairie Mosaic." In *Conservation in Highly Fragmented Landscapes*, edited by M. W. Schwartz, 55–87. New York: Chapman and Hall.

Roe, J. H., B. A. Kingsbury, and N. R. Herbert. 2003. "Wetland and Upland Use Patterns in Semi-Aquatic Snakes: Implication for Wetland Conservation." *Wetlands* 23 (4).

Samson, F. B., and F. L. Knopf. 1994. "Prairie Conservation in North America." *BioScience* 44: 418–21.

Smithsonian Institute Archives. 2011. "Spencer Fullerton Baird, 1823–1887." April 14, 2011. https://siarchives.si.edu/history/spencer-fullerton-baird.

———. 2011. "Charles Frederic Girard Papers, circa 1846–1860 and Undated." September 16, 2011. https://siarchives.si.edu/collections/siris_arc_217347.

Warner, R. E. 1991. "Farm Programs, Agricultural Technologies, and Upland Wildlife Habitat." In "Our Living Heritage: The Biological Resources of Illinois," edited by L. M. Page and M. R. Jeffords, special issue, *Illinois Natural History Survey Bulletin* 34 (4): 457.

VULNERABILITY

Barton, Christopher, and Karen Kinkead. 2005. "Do Erosion Control and Snakes Mesh?" *Journal of Soil and Water Conservation* 60 (2): 33A–35A.

Bernardino, Frank S., and George H. Dalrymple. 1992. "Seasonal Activity and Road Mortality of the Snakes of the Pa-Hay-Okee Wetlands of Everglades National Park, USA." *Biological Conservation* 62 (2): 71–75. https://doi.org/10.1016/0006-3207(92)90928-G.

Cagle, Nicolette L. 2008. "A Multiscale Investigation of Snake Habitat Relationships and Snake Conservation in Illinois." Ph.D. diss., Duke University.

Gregory, P. T., J. M. Macartney, and K. W. Larsen. 1987. "Spatial Patterns and Movements." In *Snakes Ecology and Evolutionary Biology*, edited by R. A. Seigel, J. T. Collins, and S. S. Novak, 366–95. Caldwell, NJ: Blackburn.

Harding, J. H. 1997. *Amphibians and Reptiles of the Great Lakes Region*. Ann Arbor: University of Michigan Press.

Hartmann, Paulo A., Marilia T. Hartmann, and Marcio Martins. 2011. "Snake Road Mortality in a Protected Area in the Atlantic Forest of Southeastern Brazil." *South American Journal of Herpetology* 6 (1): 35–42. https://doi .org/10.2994/057.006.0105.

Kovar, Roman, Marek Brabec, Radovan Vita, and Radomir Bocek. 2014. "Mortality Rate and Activity Patterns of an Aesculapian Snake (Zamenis longissimus) Population Divided by a Busy Road." *Journal of Herpetology* 48 (1): 24–33. https://doi.org/10.1670/12-090.

Laurance, William F., Gopalasamy Reuben Clements, Sean Sloan, Christine S. O'Connell, Nathan D. Mueller, Miriam Goosem, Oscar Venter, et al. 2014. "A Global Strategy for Road Building." *Nature* 513 (7517): 229–32. https:// doi.org/10.1038/nature13717.

Lesser, Elizabeth. 2020. *Cassandra Speaks*. New York: Harper Wave.

McDonald, P. J. 2012. "Snakes on Roads: An Arid Australian Perspective." *Journal of Arid Environments* 79 (April): 116–19. https://doi.org/10.1016/j .jaridenv.2011.11.028.

Nickens, T. Edward. 2016 "Riddle of the Bays." *Our State*, May 17, 2016. https://www.ourstate.com/riddle-of-the-bays/.

Parker, W. S., and M. V. Plummer. 1987. "Population Ecology." In *Snakes Ecology and Evolutionary Biology*, edited by R. A. Seigel, J. T. Collins, and S. S. Novak, 363–94. Caldwell, NJ: Blackburn.

Phillips, C. A., R. A. Brandon, and E. O. Moll. 1999. *Field Guide to Amphibians and Reptiles of Illinois*. Champaign: Illinois Natural History Survey.

Saint Giron, H. 1992. "Stratégies reproductives des viperidae dans les zones tempérées fraîches et froides." *Bulletin de la Société Zoologique de France*. https://pascal-francis.inist.fr/vibad/index.php?action=getRecordDetail& idt=4755447.

Shepard, Donald B., Michael J. Dreslik, Benjamin C. Jellen, and Christopher A. Phillips. "Reptile Road Mortality around an Oasis in the Illinois Corn Desert with Emphasis on the Endangered Eastern Massasauga." 2008. *Copeia* 2: 350–59.

Wikipedia, s.v. "Carolina Bays." 2019. https://en.wikipedia.org/w/index.php ?title=Carolina_bays&oldid=932317207.

Yue, Sam, Timothy C. Bonebrake, and Luke Gibson. 2019. "Informing Snake Roadkill Mitigation Strategies in Taiwan Using Citizen Science." *Journal*

of Wildlife Management 83 (1): 80–88. https://doi.org/10.1002/jwmg
.21580.

THE OUTBACK

"Biodiscovery in Queensland." 2011. September 21, 2011. https://www
.business.qld.gov.au/industries/science-it-creative/science/biodiscovery
/qld.

Bradford, Alina, and Nicoletta Lanese. 2022. "Platypus Facts." February 18,
2022. https://www.livescience.com/27572-platypus.html.

"Cane Toad." 2010. September 10, 2010. https://www.nationalgeographic
.com/animals/amphibians/c/cane-toad/.

"Common Tree Snake." n.d. http://www.wildlifeqld.com.au/common-tree
-snake.html.

Cook, Terri, and Lon Abbott. 2018. "Travels in Geology: Northeastern
Australia's Atherton Tablelands Young Volcanoes, Waterfalls and Rainforest near the Great Barrier Reef." *Earth*, June 5, 2018. https://www
.earthmagazine.org/article/travels-geology-northeastern-australias
-atherton-tablelands-young-volcanoes-waterfalls-and.

Encyclopedia Britannica, s.v. "Platypus | Eggs, Habitat, & Facts." Accessed
March 13, 2019. https://www.britannica.com/animal/platypus.

Fearn, S., L. Schwarzkopf, and R. Shine. 2005. "Giant Snakes in Tropical Forests: A Field Study of the Australian Scrub Python, Morelia kinghorni."
Wildlife Research 32 (2): 193. https://doi.org/10.1071/WR04084.

Hartley, Anna. 2018. "Snake Undergoes Surgery after Eating Girl's 'Cow.'"
July 13, 2018. https://www.abc.net.au/news/2018-07-13/snake-eats
-stuffed-toy-and-undergoes-emergency-surgery-in-cairns/9991212.

Lewis, Martin W. 2012. "The Culture of Queensland and the Desire to Divide
the State." *GeoCurrents* (blog). http://www.geocurrents.info/cultural
-geography/the-culture-of-queensland-and-the-desire-to-divide-the
-state.

Martin, W. R. 1995. "Field Observation of Predation on Bennett's Tree-
Kangaroo (Dendrolagus bennettianus) by an Amethystine Python (Morelia amethistina)." *Herpetological Review* 26 (1995): 74–76.

Phillips, Ben L, and Richard Shine. 2006. "An Invasive Species Induces Rapid
Adaptive Change in a Native Predator: Cane Toads and Black Snakes

in Australia." *Proceedings of the Royal Society B: Biological Sciences* 273 (1593): 1545–50. https://doi.org/10.1098/rspb.2006.3479.

"Platypus." 2010. September 10, 2010. https://www.nationalgeographic.com /animals/mammals/p/platypus/.

Shine, Richard. 1995. *Australian Snakes: A Natural History.* Ithaca, NY: Cornell University Press.

Wikipedia, s.v. "Dendrelaphis punctulatus." 2018. https://en.wikipedia.org/w /index.php?title=Dendrelaphis_punctulatus&oldid=875237943.

PERU

Bryan, Jenny. 2009. "From Snake Venom to ACE Inhibitor—the Discovery and Rise of Captopril." *Pharmaceutical Journal* 282 (April): 455.

Layt, Stuart. 2020. "Unique Snake Venom Discovery Could Help Put the Bite on Alzheimer's." *Brisbane Times,* February 11, 2020. https://www.brisbanetimes.com.au /national/queensland/unique-snake-venom-discovery-could -help-put-the-bite-on-alzheimer-s-20200211-p53zuu.html?fbclid =IwAR21vfBJ9xtBcyymX2WP2NcmBE86Ct53NGNL2V5TX2hFSOn0YdC _56_gTro.

Martin, Douglas. 2011. "Bill Haast, a Man Charmed by Snakes, Dies at 100." *New York Times,* June 17, 2011, sec. U.S. https://www.nytimes.com/2011 /06/18/us/18haast.html.

Mooney, Carolyn J. 1998. "In the Amazon, a Library Reflects a Mythology of River and Jungle." *Chronicle of Higher Education* 45 (17): B2. https://search-proquest-com.proxy.lib.duke.edu/docview /214741094?https://www.nclive.org/cgi-bin/nclsm?rsrc=306&pq-origsite =summon.

Roberto, Igor Joventino, and Robinson Botero-Arias. 2013. "The Distress Call of Caiman Crocodilus crocodilus (Crocodylia: Alligatoridae) in Western Amazonia, Brazil." *Zootaxa* 3647 (4): 593. https://doi.org/10.11646 /zootaxa.3647.4.9.

Tobin, John E. 1997. "Competition and Coexistence of Ants in a Small Patch of Rainforest Canopy in Peruvian Amazonia." *Journal of the New York Entomological Society* 105 (1/2): 105–12.

Vergne, A. L., M. B. Pritz, and N. Mathevon. 2009. "Acoustic Communication in Crocodilians: From Behaviour to Brain." *Biological Reviews* 84 (3): 391–411. https://doi.org/10.1111/j.1469-185X.2009.00079.x.

Wikipedia, s.v. "Bill Haast." 2020. https://en.wikipedia.org/w/index.php?title=Bill_Haast&oldid=946187759.

GUMBY

Cagle, Nicolette L. 2008. "A Multiscale Investigation of Snake Habitat Relationships and Snake Conservation in Illinois." Ph.D. diss., Duke University.

Carpenter, Charles C. 1982. "The Bullsnake as an Excavator." *Journal of Herpetology* 16 (4): 394. https://doi.org/10.2307/1563570.

Conant, R., and J. T. Collins. 1991. *A Field Guide to Reptiles and Amphibians: Eastern and Central North America.* Boston: Houghton Mifflin.

Harding, J. H. 1997. *Amphibians and Reptiles of the Great Lakes Region.* Ann Arbor: University of Michigan Press.

Kissner, K. J., and J. Nicholson. 2003. *Bullsnakes (Pituophis catenifer sayi) in Alberta: Literature Review and Data Compilation.* Alberta Sustainable Resource Department, Fish and Wildlife Division, Species at Risk Report no. 62.

"Leigh Farm." Accessed March 14, 2019. http://www.opendurham.org/buildings/leigh-farm.

Phillips, C. A., R. A. Brandon, and E. O. Moll. 1999. *Field Guide to Amphibians and Reptiles of Illinois.* Champaign: Illinois Natural History Survey.

THE NEXT GENERATION

Calderon-Arrieta, Diego. 2017. "Evaluating Current Attitudes towards Snakes in the Nicholas School of the Environment's (NSOE) Environmental Master's Student Community." Master's thesis, Duke University, Nicholas School of the Environment.

Gerald, Gary W., Mark J. Mackey, and Dennis L. Claussen. 2008. "Effects of Temperature and Perch Diameter on Arboreal Locomotion in the Snake Elaphe guttata." *Journal of Experimental Zoology Part A: Ecological Genetics and Physiology* 309A (3): 147–56. https://doi.org/10.1002/jez.443.

Gill, Nicholas Michael. 2017. "Near Eastern Snake Omens and Roman Literature." Master's thesis, Queen's University, Kingston, Ontario. https://qspace.library.queensu.ca/bitstream/handle/1974/22022/Gill_Nicholas_M_201708_MA.pdf?sequence=1.

Goldsmith, S. K. 1988. "Courtship Behavior of the Rough Green Snake, Opheodrys aestivus (Colubridae: Serpentes)." *Southwestern Naturalist* 33 (4): 473. https://doi.org/10.2307/3672215.

Guyer, Craig, and Maureen A. Donnelly. 1990. "Length-Mass Relationships among an Assemblage of Tropical Snakes in Costa Rica." *Journal of Tropical Ecology* 6 (1): 65–76.

Hawkins, Roxanne D., and Joanne M. Williams. 2017. "Childhood Attachment to Pets: Associations between Pet Attachment, Attitudes to Animals, Compassion, and Humane Behaviour." *International Journal of Environmental Research and Public Health* 14 (5). https://doi.org/10.3390/ijerph14050490.

Joyner, Kelly, and Hannah Royal. 2020. "Management for an Imperiled Reptile on a Barrier Island: Eastern Diamondback Rattlesnake (Crotalus adamanteus)." Master's thesis, Duke University, Nicholas School of the Environment, Durham, NC. https://dukespace.lib.duke.edu/dspace/handle/10161/20480.

Lewbart, G. A., J. Kishimorig, and L. S. Christian. 2010. "The North Carolina State University College of Veterinary Medicine Turtle Rescue Team: A Model for a Successful Wild-Reptile Clinic." *Journal of Wildlife Rehabilitation* 30 (1): 25–30.

Miao, Ruolin E., and Nicolette L. Cagle. 2020. "The Role of Gender, Race, and Ethnicity in Environmental Identity Development in Undergraduate Student Narratives." *Environmental Education Research* 26 (2): 171–88. https://doi.org/10.1080/13504622.2020.1717449.

"NCpedia." Accessed November 3, 2020. https://www.ncpedia.org/anchor/naval-stores-and-longleaf.

Özel, Murat, Pavol Prokop, and Muhammet Uşak. 2009. "Cross-Cultural Comparison of Student Attitudes toward Snakes." *Society & Animals* 17 (3): 224–40. https://doi.org/10.1163/156853009X445398.

Palmer, William M., and Alvin L. Braswell. 1976. "Communal Egg Laying and Hatchlings of the Rough Green Snake, Opheodrys aestivus (Linnaeus) (Reptilia, Serpentes, Colubridae)." *Journal of Herpetology* 10 (3): 257. https://doi.org/10.2307/1562991.

Plummer, Michael V. 1990. "High Predation on Green Snakes, Opheodrys aestivus." *Journal of Herpetology* 24 (3): 327. https://doi.org/10.2307/1564409.

Sisson, Bryan. 2013. *Peribology: A Budding of Secrets Compiling Cultural Insights Regarding Nature's Treasures.* Bloomington, IN: Xlibris.

Steen, David. 2011. "Cottonmouth Myths I: Snakes Dropping into Boats." *Living alongside Wildlife* (blog). January 15, 2011. http://www.livingalongsidewildlife.com/2011/01/cottonmouth-myths-i-snakes-dropping.html.

Young, Ashley, Kathayoon A. Khalil, and Jim Wharton. 2018. "Empathy for Animals: A Review of the Existing Literature." *Curator: The Museum Journal* 61 (2): 327–43. https://doi.org/10.1111/cura.12257.

WORDS AND WISDOM

Beane, Jeffrey C., Sean P. Graham, Thomas J. Thorp, and L. Todd Pusser. 2014. "Natural History of the Southern Hognose Snake (Heterodon Simus) in North Carolina, USA." 2014. *Copeia* 1: 168–75. https://doi.org/10.1643/CH-13-044.

Gibbons, J. Whitfield, and Raymond D. Semlitsch. 1987. "Activity Patterns." In *Snakes: Ecology and Evolutionary Biology*, 396–421. Caldwell, NJ: Blackburn.

Nerburn, K. 2002. *Neither Wolf nor Dog: On Forgotten Roads with an Indian Elder.* Novato, CA: New World Library.

THE MIRROR

Giesen, James C. 2020. "The View from Rose Hill: Environmental, Architectural, and Cultural Recovery on a Piedmont Landscape." *Buildings & Landscapes: Journal of the Vernacular Architecture Forum* 27 (2): 19–38. https://doi.org/10.5749/buildland.27.2.0019.

Jefferson, T. 1797. "A Memoir on the Discovery of Certain Bones of a Quadruped of the Clawed Kind in the Western Parts of Virginia." *Transactions of the American Philosophical Society* 4: 246–60.

Leopold, Aldo. 1992. *The River of the Mother of God and Other Essays by Aldo Leopold.* Madison: University of Wisconsin Press.

Roy, Sree. 2011. "S-s-snakeskin mani-pedi." August 12, 2011. www.nailsmag
.com/demoarticle/93308/s-s-snakeskin-mani-pedi.

Wilder, Thornton. 2003. *Our Town.* New York: Harper Perennial.

DISAFFECTED

"The Rattlesnake as a Symbol of America—by Benjamin Franklin." n.d.
http://greatseal.com/symbols/rattlesnake.html.

"To People Who Drive with 'Don't Tread on Me' Flag License Plates." n.d.
NeoGAF. Accessed August 29, 2019. https://www.neogaf.com/threads/to
-people-who-drive-with-dont-tread-on-me-flag-license-plates.1191440/.

Walker, Rob. 2016. "The Shifting Symbolism of the Gadsden Flag." *New
Yorker,* October 2, 2016. https://www.newyorker.com/news/news-desk
/the-shifting-symbolism-of-the-gadsden-flag.

Wikipedia, s.v. "Rod of Asclepius." 2019. https://en.wikipedia.org/w/index
.php?title=Rod_of_Asclepius&oldid=910462227.

A *JUBO* IN CUBA

Jaume, Miguel L., and Orlando H. Garrido. 1980. "Notas Sobre Mordeduras
de Jubo Alsophis cantherigerus Bibron (Reptilia-Serpentes Colubridae)
En Cuba." *Revista Cubana de Medicina Tropical* 32 (2): 145–48.

León, Sharon, and David Huerta. 2015. *Semillas del canto: Mujeres latino-
americanas en la poesía.* Mexico City: Ediciones SM.

Neill, Wilfred T. 1954. "Evidence of Venom in Snakes of the Genera Alsophis
and Rhadinaea." *Copeia* 1954 (1): 59–60.

Rodríguez Schettino, Lourdes, Julio Larramendi, and Instituto de Ecologia y
Sistemática (Cuba). 2003. *Anfibios y reptiles de Cuba.* Vasa, Finland: UPC
Print.

HOPE FOR THE FUTURE

Ehrenfeld, Joan G., and David W. Ehrenfeld. 1973. "Externally Secreting
Glands of Freshwater and Sea Turtles." *Copeia* 1973 (2): 305–14. https://
doi.org/10.2307/1442969.

Lewis, C. H., S. F. Molloy, R. M. Chambers, and J. Davenport. 2007. "Response of Common Musk Turtles (Sternotherus odoratus) to Intraspecific Chemical Cues." *Journal of Herpetology* 41 (3): 349–53.

Martof, Bernard S., William M. Palmer, Joseph F. Bailey, and Julian R. Harrison. 1980. *Amphibians and Reptiles of the Carolinas and Virginia.* Chapel Hill: University of North Carolina Press.

Rohde, Fred C., Rudolf G. Arndt, David G. Lindquist, and James F. Parnell. 1994. *Freshwater Fishes of the Carolinas, Virginia, Maryland, and Delaware.* Chapel Hill: University of North Carolina Press.

INDEX

Australia (continued)
    Kakadu National Park; King's
    Canyon; Northern Territory;
    Queensland; Tablelands; Yung-
    aburra
Australian Taipan (Oxyuranus scutel-
    latus scutellatus), 10

Baez, Firelei, 34
Baird, Spencer, 69–70; natural his-
    tory collection of, 69
Balkans, 5
Ball Python (Python regius), 57. See
    also pythons
Banksy, 34
Barrine, Lake, 97
bats: declines of species of, 87; of
    Guam, 37. See also bat species
bat species: Fishing Bat (Noctilio
    leporinus), 113; Flying Fox, 95;
    Little Mariana Fruit Bat (Ptero-
    pus mariannus), 37; Proboscis Bat
    (Rhynchonycteris naso), 40
Beane, Jeff, 140
beauty: of snakes, 2, 34–36, 43, 49,
    53, 72–73, 79, 103, 118; of snakes in
    works of art, 33–34
Bennett's Tree Kangaroo (Dendrola-
    gus bennettianus), 99
Bible, 35, 41–42
billabongs, 94
bimodal peak of activity, 140
biodiversity, 83, 95, 112, 150
birds: in captivity, 37; of Guam, 37; of
    North Carolina Birding Trail, 14
bird species: Acadian Flycatcher
    (Empidonax virescens), 140;

Aracari (Pteroglossus species),
    113; Arctic Warbler (Phylloscopus
    borealis kennicottii), 22; Bald
    Eagle (Haliaeetus leucocephalus),
    24, 156, 158; Blue-crowned Mot-
    mot (Momotus coeruliceps), 41;
    Blue-gray Gnatcatcher (Polioptila
    caerulea), 170; Blue Jay (Cyanoc-
    itta cristata), 133, 171; Burrowing
    Owls (Athene cunicularia), 60;
    Carolina Chickadee (Poecile
    carolinensis), 166; Carolina Wren
    (Thryothorus ludovicianus), 84;
    Eastern Screech Owl (Megas-
    cops asio), 30; Fish Crow (Corvus
    ossifragus), 171; Great-crested
    Flycatcher (Myiarchus crinitus),
    84; Great Egret (Ardea alba),
    102; Guam Flycatcher (My-
    iagra freycineti), 37; Guam Rail
    (Ko'ko') (Hypotaenidia owstoni),
    37; hawks, 2, 143, 148; Horned
    Screamer (Anhima cornuta),
    102, 113; Jacana (Jacana jacana),
    102; Little Penguin (Eudyptula
    minor), 51; lorikeet, 94; Northern
    Cardinal (Cardinalis cardinalis),
    84, 166; Northern Parula (Seto-
    phaga americana), 78; Oropendola
    (Psarocolius species), 113; Red-
    bellied Woodpecker (Melanerpes
    carolinus), 84; Red-cockaded
    Woodpecker (Picoides borealis),
    132; Spinifex Pigeon (Geophaps
    plumifera), 91–92; Spotted Shag
    (Phalacrocorax punctatus), 51;
    Toucan (family Ramphastidae),

flua), 83, 164; Tulip Tree (*Liriodendron tulipifera*), 154; Water Oak (*Quercus nigra*), 164; White Pine (*Pinus strobus*), 135

Turkey, 17

turtle species: Alligator Snapping Turtle (*Macrochelys temminckii*), 31–32; Eastern Box Turtle (*Terrapene carolina carolina*), 152; Eastern Musk Turtle (*Sternotherus odoratus*), 170; Saw-shelled Terrapin (*Myuchelys latisternum*), 98; Slider Turtle (*Trachemys* genus), 167

Tuscaroras, 153. *See also* indigenous peoples

Tutelo, 116, 153. *See also* indigenous peoples

Ugajin, 33

Uluru (Ayer's Rock), 91

unimodal peak of activity, 140

United Kingdom, 56–57, 135

United States: northeastern, 135; northern, 83; southeastern, 42, 135; southern, 135; western, 68. *See also individual states*

United States Army Ammunition Plant, 67

United States Department of Agriculture, 38

United States Endangered Species Act, 23

United States Forest Service, 67, 149

upland forest, 164

urbanization, 23, 73, 116, 126

Uruguay, 59

veldt, 59

Vence, France, 12–14

venom: chemistry of, 107; cytotoxic, 107; enzymes in, 105–7; experiments on the effects of rattlesnake, 19; extraction of, 111; glands of, 63–64; hemotoxic, 107; proteolytic, 107; research on pharmaceutical properties of, 10, 111. *See also* snake bites

Vermont, 158

viperids, 10. *See also* pit vipers

Virginia, 12, 83, 152, 158; historic sites of, 153; Jefferson's list of the birds of, 17. *See also* Fort Christanna; Mount Vernon

Waccamaw, Lake, 78–79

Walker, Rob, 156–57

wallaby, 94

Washington, DC, 152, 155

Washington, George, 86, 153–55

wasichu (white people), 143

watersnakes (*Nerodia* species), 15, 32–33, 53, 82, 127. *See also Nerodia erythrogaster* (Red-bellied Watersnake); *Nerodia erythrogaster flavigaster* (Yellow-bellied Watersnake); *Nerodia sipedon* (Northern Watersnake); *Nerodia sipedon pleuralis* (Midland Watersnake)

Weber, Andreas, 32

Western Diamondback Rattlesnake (*Crotalus atrox*), 27

Western Fox Snake (*Pantherophis ramspotti*), 70, 72. *See also* fox snakes

Printed in the USA
CPSIA information can be obtained
at www.ICGtesting.com
CBHW020027210524
8864CB00001B/16